SCIENCE
A CLOSER LOOK

BUILDING SKILLS

Reading and Writing

Mc Graw Hill Macmillan McGraw-Hill

Instructions for Copying

Answers are printed in non-reproducible blue. Copy pages on a light setting in order to make multiple copies for classroom use.

Contents

LIFE SCIENCE

Unit A Literature 1

Chapter 1 Plants
Chapter Concept Map. 2
Lesson 1 Lesson Outline 3
          Lesson Vocabulary 5
          Lesson Cloze Activity 6
Lesson 2 Lesson Outline 7
          Lesson Vocabulary 9
          Lesson Cloze Activity 10
          Writing in Science 11
Lesson 3 Lesson Outline 13
          Lesson Vocabulary 15
          Lesson Cloze Activity 16
          Reading in Science 17
Chapter Vocabulary 19

Chapter 2 Animals
Chapter Concept Map. 21
Lesson 1 Lesson Outline 22
          Lesson Vocabulary 24
          Lesson Cloze Activity 25
Lesson 2 Lesson Outline 26
          Lesson Vocabulary 28
          Lesson Cloze Activity 29
          Reading in Science 30
Lesson 3 Lesson Outline 32
          Lesson Vocabulary 34
          Lesson Cloze Activity 35
          Writing in Science 36
Chapter Vocabulary 38

Contents

Unit B Literature 40

Chapter 3   Looking at Habitats

Chapter Concept Map. 41

Lesson 1 Lesson Outline 42

Lesson Vocabulary 44

Lesson Cloze Activity 45

Lesson 2 Lesson Outline 46

Lesson Vocabulary 48

Lesson Cloze Activity 49

Writing in Science 50

Lesson 3 Lesson Outline 52

Lesson Vocabulary 54

Lesson Cloze Activity 55

Reading in Science 56

Chapter Vocabulary 58

Chapter 4   Kinds of Habitats

Chapter Concept Map. 60

Lesson 1 Lesson Outline 61

Lesson Vocabulary 63

Lesson Cloze Activity 64

Reading in Science 65

Lesson 2 Lesson Outline 67

Lesson Vocabulary 69

Lesson Cloze Activity 70

Lesson 3 Lesson Outline 71

Lesson Vocabulary 73

Lesson Cloze Activity 74

Writing in Science 75

Chapter Vocabulary 77

Contents

EARTH SCIENCE

Unit C Literature 79

Chapter 5   Land and Water

Chapter Concept Map 80

Lesson 1   Lesson Outline 81

              Lesson Vocabulary 83

              Lesson Cloze Activity 84

Lesson 2   Lesson Outline 85

              Lesson Vocabulary 87

              Lesson Cloze Activity 88

              Writing in Science 89

Lesson 3   Lesson Outline 91

              Lesson Vocabulary 93

              Lesson Cloze Activity 94

              Reading in Science 95

Chapter Vocabulary 97

Chapter 6   Earth's Resources

Chapter Concept Map 99

Lesson 1   Lesson Outline 100

              Lesson Vocabulary 102

              Lesson Cloze Activity 103

              Writing in Science 104

Lesson 2   Lesson Outline 106

              Lesson Vocabulary 108

              Lesson Cloze Activity 109

Lesson 3   Lesson Outline 110

              Lesson Vocabulary 112

              Lesson Cloze Activity 113

              Reading in Science 114

Chapter Vocabulary 116

Contents

Unit D Literature. 118

Chapter 7 Observing Weather

Chapter Concept Map. 119

Lesson 1 Lesson Outline. 120

Lesson Vocabulary 122

Lesson Cloze Activity 123

Writing in Science 124

Lesson 2 Lesson Outline. 126

Lesson Vocabulary 128

Lesson Cloze Activity 130

Lesson 3 Lesson Outline. 131

Lesson Vocabulary 132

Lesson Cloze Activity 133

Reading in Science 134

Chapter Vocabulary 136

Chapter 8 Earth and Space

Chapter Concept Map. 138

Lesson 1 Lesson Outline. 139

Lesson Vocabulary 141

Lesson Cloze Activity 142

Lesson 2 Lesson Outline. 143

Lesson Vocabulary 145

Lesson Cloze Activity 146

Writing in Science 147

Lesson 3 Lesson Outline. 149

Lesson Vocabulary 151

Lesson Cloze Activity 152

Lesson 4 Lesson Outline. 153

Lesson Vocabulary 155

Lesson Cloze Activity. 156

Reading in Science 157

Chapter Vocabulary 159

Contents

PHYSICAL SCIENCE

Unit E Literature. 161

Chapter 9 Looking at Matter

Chapter Concept Map. 162

Lesson 1 Lesson Outline. 163

Lesson Vocabulary .165

Lesson Cloze Activity. 166

Lesson 2 Lesson Outline. 167

Lesson Vocabulary .169

Lesson Cloze Activity. 170

Reading in Science .171

Lesson 3 Lesson Outline. 173

Lesson Vocabulary .175

Lesson Cloze Activity. 176

Writing in Science 177

Chapter Vocabulary 179

Chapter 10 Changes in Matter

Chapter Concept Map. 181

Lesson 1 Lesson Outline. 182

Lesson Vocabulary .184

Lesson Cloze Activity. 185

Lesson 2 Lesson Outline. 186

Lesson Vocabulary .188

Lesson Cloze Activity. 189

Reading in Science .190

Lesson 3 Lesson Outline. 192

Lesson Vocabulary .194

Lesson Cloze Activity. 195

Writing in Science 196

Chapter Vocabulary 198

Contents

Unit F Literature. 200

Chapter 11 How Things Move

Chapter Concept Map. 201

Lesson 1 Lesson Outline 202

Lesson Vocabulary 204

Lesson Cloze Activity 205

Lesson 2 Lesson Outline 206

Lesson Vocabulary 208

Lesson Cloze Activity 209

Reading in Science 210

Lesson 3 Lesson Outline 212

Lesson Vocabulary 214

Lesson Cloze Activity 215

Writing in Science 216

Lesson 4 Lesson Outline 218

Lesson Vocabulary 220

Lesson Cloze Activity 221

Chapter Vocabulary 222

Chapter 12 Using Energy

Chapter Concept Map. 224

Lesson 1 Lesson Outline 225

Lesson Vocabulary 227

Lesson Cloze Activity 228

Lesson 2 Lesson Outline 229

Lesson Vocabulary 231

Lesson Cloze Activity 232

Writing in Science 233

Lesson 3 Lesson Outline 235

Lesson Vocabulary 237

Lesson Cloze Activity 238

Lesson 4 Lesson Outline 239

Lesson Vocabulary 241

Lesson Cloze Activity 242

Reading in Science 243

Chapter Vocabulary 245

Name _____ Date _____

The Seed

by Aileen Fisher

Read the Unit Literature pages in your book.

 Write About It

Response to Literature

1. What do you think seeds need to grow?

Possible answer: I think seeds need water and

sunlight to grow.

2. Where have you seen seeds? Draw a picture.

Drawings will vary.

Plants

Fill in the plant parts as you read the chapter.

flower	leaves	seed
fruit	roots	stem

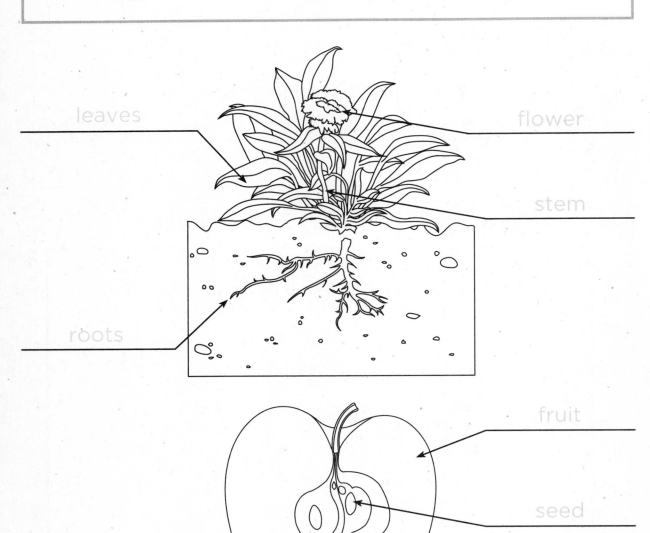

leaves

flower

stem

roots

fruit

seed

What Living Things Need

Use your book to help you fill in the blanks.

What do living things need?

1. All living things have needs they must meet in

   order to grow and _____change_____ .

2. Most animals need to move, _____breathe_____
   air, drink water, and eat food to grow.

3. Plants are _____living_____ things, too.

4. Plants also change and _____grow_____ over
   time.

5. Plants need _____air_____ , water, and space
   to grow.

6. Plants use their parts to make their own
   _____food_____ .

Name _____ Date _____

How do plants make food?

7. Plants use ___sunlight___ , air, water, and minerals to make their own food.

8. Minerals come from tiny bits of ___soil___ and rocks.

9. When plants make their own food, they also make a ___gas___ called oxygen.

10. People and animals need ___oxygen___ to breathe.

Critical Thinking

11. How do the parts of a plant help it get what it needs to live?

Possible answer: The leaves of a plant help it take in air and sunlight. Food and water go through the plant's stem to get to all parts of the plant. The roots of the plant take in water and minerals.

What Living Things Need

Choose a word from the box to answer each riddle.

leaves	oxygen	stem
minerals	roots	

1. I can be found in the ground. I am made from tiny bits of rocks and soil. What am I?

minerals

2. I help a plant take in air and sunlight. What am I?

leaves

3. Food and water travel through me to get to all parts of the plant. What am I?

stem

4. I help a plant take in minerals from the ground. What am I?

roots

5. You breathe me in so you can live. What am I?

oxygen

Name _____ Date _____

What Living Things Need

Fill in the blanks. Use the words from the box.

gas	minerals	roots	sunlight
leaves	oxygen	stem	

Plants, animals, and people all need food, air, and water to live. Plants need __sunlight__ and space to grow, too.

Animals and people must move around to get what they need, but plants have parts that help them live in their environments. The __roots__ hold the plant in the ground. They also take in __minerals__ from soil.

Food and water travel through the __stem__ to reach all parts of the plant. The __leaves__ take in air and sunlight to make food. When plants make food, they give off __oxygen__ into the air.

Oxygen is a __gas__ that helps us breathe. We can find oxygen in the air we breathe and the water we drink.

© Macmillan/McGraw-Hill

Plants Make New Plants

Use your book to help you fill in the blanks.

Where do seeds come from?

1. A _____seed_____ is a special plant part that can grow into a new plant.

2. Seeds are made inside a _____flower_____ .

3. Sometimes a flower will _____grow_____ seeds inside of a fruit.

4. Flowers also make _____pollen_____ , the sticky powder that helps them make seeds.

5. Bugs and _____birds_____ can help move pollen from flower to flower.

6. Wind and _____water_____ from rain can move pollen, too.

How do seeds look?

7. Seeds can have many _____sizes_____ and shapes, just like plants.

8. All seeds have seed _____coats_____ or fruit to protect them as they grow.

How do seeds grow?

9. The _____life cycle_____ of a plant begins with a seed.

10. The way plants grow, live, and _____die_____ is called their life cycle.

11. Most seeds need _____light_____ , water, food, and a little heat to become new plants.

12. A new plant has the same life cycle as its _____parent_____ plant.

Critical Thinking

13. How are new plants that grow from seeds like their parent plants?

Possible answer: New plants that grow from seeds have the same features and life cycles as their parent plants.

© Macmillan/McGraw-Hill

Plants Make New Plants

Read the sentences below. Write TRUE if the sentence is true. Write NOT TRUE if the sentence is false.

1. _NOT TRUE_____ Inside a seed, there is a sticky powder called pollen.

2. _TRUE_____ Part of a flower can turn into fruit.

3. _TRUE_____ The fruit protects the seeds inside it.

4. _TRUE_____ A life cycle shows how a plant grows, lives, and dies.

5. _NOT TRUE_____ An adult plant can grow into a seedling.

6. _TRUE_____ Seeds have a special coat that keeps them from drying out.

Name _____ Date _____

Plants Make New Plants

Fill in the blanks. Use the words from the box.

flowers	life cycle	seed coat	seeds
fruit	pollen	seedling	

Plants make new plants during their life cycle. A

_____life cyle_____ shows how a living thing grows,

lives, and dies. The life cycle of a plant begins with

a seed. A special covering called a ___seed coat___

helps protect the seed. The seed sprouts a ___seedling___

if it gets enough food, water, and heat. It may grow

___flowers___ as it becomes an adult plant.

A sticky material called ___pollen___ is found

inside of flowers. Flowers use pollen to make seeds.

Part of the flower can also grow into a fruit that has

___seeds___ . When the ___fruit___

becomes ripe, it falls to the ground. Then the seeds

can turn into new plants.

Main Idea and Details

✏️ **Write About It**

On a separate piece of paper, write a paragraph about a flower that you observed. Include a main idea and details.

Getting Ideas

Write the name of a flower in the Main Idea oval. Write a detail about the flower in each detail oval.

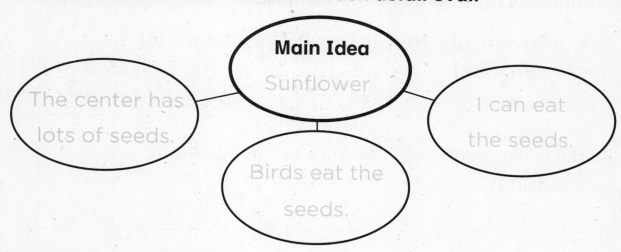

Main Idea

Sunflower

The center has lots of seeds.

I can eat the seeds.

Birds eat the seeds.

Planning and Organizing

Hector wrote three sentences about a sunflower. Write Detail if the sentence tells a detail. Write Main Idea if the sentence tells the main idea.

1. _____Detail_____ Birds like sunflower seeds.

2. _____Detail_____ A sunflower has seeds.

3. ___Main Idea___ A sunflower is useful.

Drafting

Write a sentence that tells the main idea about your flower.

Possible sentence: Sunflowers are very useful.

On a separate piece of paper, write a whole paragraph. Give details about your flower.

Revising and Proofreading

Hector wrote some sentences. Use the words in parentheses ( ) to combine his sentences.

1. Sunflowers are easy to grow. They need a lot of room. (but)

   Sunflowers are easy to grow, but they need a lot of room.

2. Birds like sunflower seeds. People like them, too. (and)

   Birds like sunflower seeds, and people like them, too.

3. The seeds are very healthful. They make a good snack. (so)

   The seeds are very healthful, so they make a good snack.

Now revise and proofread your writing. Ask yourself:

► Did I include the main ideas and details?

► Did I correct all mistakes?

How Plants Are Alike and Different

Use your book to help you fill in the blanks.

How are plants like their parents?

1. Oak trees make _____acorns_____ that grow into new oak trees.

2. Sunflowers make seeds that grow to look just like their _____parent_____ sunflower.

3. A _____trait_____ is a way a living thing looks or acts like its parent.

4. Some plants and animals share many _____traits_____ with their parents.

5. Other plants and animals _____share_____ just a few traits with their parents.

How can plants change to fit their environment?

6. An _____environment_____ is where a plant or animal lives.

7. Plants can _____change_____ to get what they need from their environment.

8. Some plants in forests grow large _____leaves_____ that help them get more sunlight.

9. Plants that live in dry places grow thick _____stems_____ to store water.

10. Plants can change to stay _____safe_____ in their environment, too.

11. Some plants grow in ways that keep away _____animals_____ that want to eat them.

12. Other plants change to stay safe from _____weather_____ where they live.

Critical Thinking

13. What do you think would happen to a plant that did not change to fit in its environment? Why?

Possible answer: I think the plant would die if it did

not change to fit in its environment. I think so because

if the plant did not change, it might not be able to get

the food, water, or sunlight it would need in order to

grow.

How Plants Are Alike and Different

Write the correct word for each sentence. Then find and circle the word in the puzzle below.

1. The ways plants and animals look and act like

   their parents are called _____traits_____ .

2. Plants can change to fit their _____environment_____ .

3. When a seed sprouts, the _____roots_____
   always grow down.

4. Plants do not pass some traits down to their

   _____offspring_____ .

E	N	V	I	R	O	N	M	E	N	T
U	S	L	H	B	F	X	F	Q	B	N
L	F	M	W	O	S	D	V	L	S	U
M	Y	E	S	L	T	R	A	I	T	S
R	O	O	T	S	H	O	W	E	K	N
A	X	A	B	K	J	L	Z	N	E	Y
B	Z	O	F	F	S	P	R	I	N	G

© Macmillan/McGraw-Hill

Name _____ Date _____

How Plants Are Alike and Different

Fill in the blanks. Use the words from the box.

dry	offspring	touches	safe
environment	parents	trait	

Some people in your family probably look alike.
They may even act alike! Plants can look and act

like their ____parents____ , too. A _____trait_____

is a way a plant or animal looks or acts like its parent.
One kind of trait that a plant could share with its

____offspring____ is the shape of its leaves.

All plants are also alike in that they can change to

fit their ____environment____ . This is where a plant lives.

Sometimes plants change in order to stay ____safe____

from the weather. Plants that live in ____dry____
places can store water in their thick stems. Plants may
also change to stay safe from animals. Some plants

can even change when an animal ____touches____

them! Venus flytraps are plants that trap and eat bugs
that wander onto their leaves.

© Macmillan/McGraw-Hill

Use with **Lesson 3**
How Plants Are Alike and Different

The Power of Periwinkle

**Read the Reading in Science pages in your book.
Use what you read to make inferences based on the
sentences in the "What I Know" column. Write your
inferences on the chart.**

What I Know	What I Infer
People who live in forests all over the world know about helpful plants.	Scientists will ask people who live in forests about helpful plants.
The rosy periwinkle was first found in the forests of Madagascar.	Scientists will look for rosy periwinkle in other forests nearby.
Scientists study plants in forests all over the world.	Scientists will find new helpful plants.

1. What did you learn about how people in Madagascar use rosy periwinkle?

   Possible answer: I learned that some people in

   Madagascar use rosy periwinkle to help cure sore

   throats.

2. Look at the picture of rosy periwinkle in your book. Draw a picture of it. Then write your own caption.

Caption: Possible answer: Rosy periwinkle is a

helpful plant that grows in the forests of Madagascar.

3. Some plants in the forest grow large _____leaves_____ that help them get more sunlight.

Write About It

Predict. What might happen if scientists find more helpful plants in the forests of the world?

Possible answer: I predict that scientists would study

the plants and try to use them to make more medicines

and help more people.

Plants

Fill in the blanks. Write the words in the puzzle.

Down

1. The reason plants and animals act like their

parents is because of ____traits____ .

3. The sticky powder inside a flower is called ____pollen____ .

5. A ____seed____ is the part of a plant that can grow into a new plant.

Across

2. The ____stem____ holds up the plant.

4. When plants make food, they give off ____oxygen____ .

1. t					**5.** s		
r		**3.** p			e		
a		**4.** o	x	y	g	e	n
i		l			d		
t		l					
2. s	t	e	m				
		n					

Name _____ Date _____

Match the words in the box to the pictures below.

| flower | leaves | roots | seedling |

I.

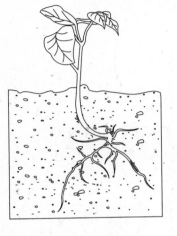

seedling
_____

2.

flower
_____

3.

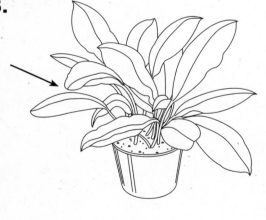

leaves
_____

4.

roots
_____

Animals

Fill in the important ideas as you read the chapter.
Some ideas have already been filled in for you.

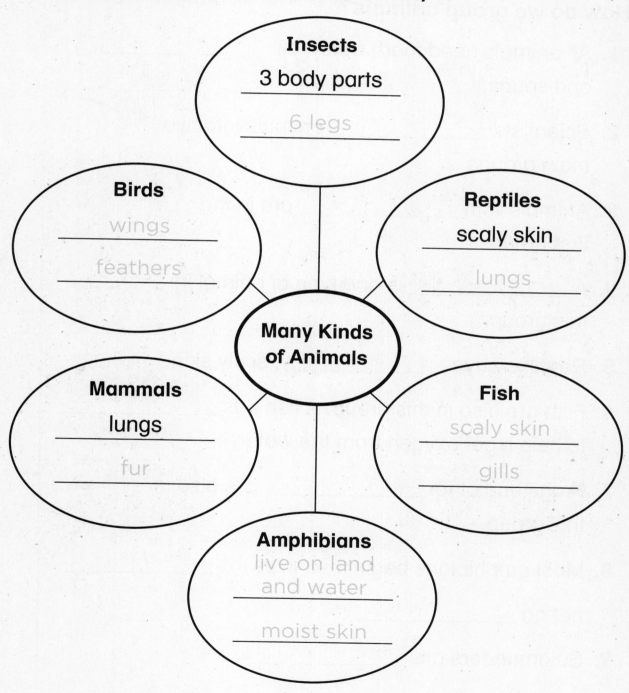

Insects
3 body parts
6 legs

Birds
wings
feathers

Reptiles
scaly skin
lungs

Many Kinds of Animals

Mammals
lungs
fur

Fish
scaly skin
gills

Amphibians
live on land and water
moist skin

© Macmillan/McGraw-Hill

Name _____ Date _____

Animal Groups

Use your book to help you fill in the blanks.

How do we group animals?

1. All animals need food, water, air, ___shelter___,
   and space.

2. Scientists ___classify___ animals into two
   main groups.

3. Animals with ___backbones___ are in the
   first group.

4. A ___reptile___ is one type of animal in
   this group.

5. Reptiles have ___rough___, scaly skin.

6. Fish are also in this group. A fish has ___gills___
   to help it get oxygen from the water.

7. Frogs and other ___amphibians___ are also in
   this group.

8. Most amphibians begin their lives in ___water___,
   not on ___land___.

9. Salamanders are ___amphibians___.

10. Birds and _____mammals_____ have backbones.

11. Mammals have _____fur_____ or hair, and birds have feathers.

What are some animals without backbones?

12. Some animals without backbones grow

coverings like _____shells_____ to keep them safe.

13. Insects have _____three_____ body parts, six legs, and no backbone.

14. The _____antennae_____ of an insect help it to feel, taste, and smell.

Critical Thinking

15. How are a bird and an insect alike? How are they different?

Birds and insects both have wings, and most of them

can fly. Birds and insects both lay eggs. Birds have a

backbone and insects do not. Birds have feathers, and

insects have hard body coverings.

_____

_____

Name _____ Date _____

Animal Groups

Label each animal with its animal group. Use the
words in the box.

amphibian	fish	mammal
bird	insect	reptile

1.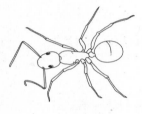

insect
_____

4.

mammal
_____

2.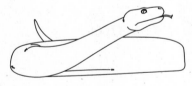

reptile
_____

5.

amphibian
_____

3.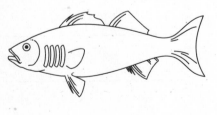

fish
_____

6.

bird
_____

© Macmillan/McGraw-Hill

Animal Groups

Fill in the blanks. Use the words from the box.

amphibian	bird	fish	mammal
backbone	classify	insect	reptile

Our world is home to many kinds of animals.

When scientists study animals, they _____classify_____

them into two groups. The groups are animals with a

_____backbone_____ and animals without a backbone.

Birds, _____fish_____ , mammals, reptiles, and

amphibians all have a backbone.

A _____bird_____ is the only animal that has

feathers. All birds have two wings, but not all birds

can fly. Fish and young amphibians use _____gills_____

to help them get oxygen from the water where they live.

An _____amphibian_____ has moist skin to help it live on

land and in water. A _____reptile_____ has dry, scaly

skin to protect it and keep it warm. A _____mammal_____

has fur and hair to keep it warm. Reptiles and mammals

use their lungs to get oxygen.

© Macmillan/McGraw-Hill

Name _____ Date _____

Animals Grow and Change

Use your book to help you fill in the blanks.

What is a life cycle?

1. A _____ life cycle _____ tells how an animal begins life, lives, and dies.

2. Insects, birds, fish, reptiles, and _____ amphibians _____ start their life cycle as eggs.

3. The life cycle of a _____ mammal _____ starts when it is born as a _____ live _____ baby.

4. Many animals look like their _____ parents _____ when they are young.

What are some other animal life cycles?

5. Some _____ animals _____ do not look like their parents at all when they are young.

6. Animals such as butterflies, frogs, and _____ crabs _____ change during their lives.

7. A caterpillar is the _____ larva _____ that hatches from a butterfly egg.

8. A caterpillar enters the _____pupa_____ stage when it is time to turn into a butterfly.

9. During this stage, the caterpillar's _____skin_____ becomes a hard shell.

10. Soon, an adult _____butterfly_____ comes out of the shell and flies away.

Critical Thinking

11. How does a human change during its life cycle?

Possible answer: First, a baby is born. It relies on
_____
its parents to get food. Then it grows bigger and
_____
stronger and learns to walk and talk. The human goes
_____
through many changes as he or she gets older.
_____

_____

_____

_____

_____

_____

_____

Animals Grow and Change

Write the correct word next to each stage of this butterfly's life cycle.

butterfly	larva
egg	pupa

1.

This animal begins as

an _____ egg _____ .

2.

When it hatches, a

_____ larva _____ comes

out. This is called a

caterpillar.

3.

The caterpillar's skin
becomes a hard shell.
This is called the

_____ pupa _____ stage.

4.

Soon, an adult

_____ butterfly _____ comes

out of the shell.

Animals Grow and Change

Fill in the blanks. Use the words from the box.

butterfly	larva	mammals	pupa
egg	life cycle	older	shell

Animals begin their lives in different ways. A _____life cycle_____ shows how an animal starts life, grows to be an adult, and dies.

Most _____mammals_____ begin their lives when they are born as live young. As they grow _____older_____, they look more like their parents.

Many insects begin life differently. A _____butterfly_____ begins life as an egg. When the _____egg_____ hatches, a _____larva_____ comes out. Soon, the larva stops moving and grows a hard _____shell_____ around its body. This is called the _____pupa_____ stage. Finally, a colorful butterfly comes out. It waits for its wings to dry and then flies away.

© Macmillan/McGraw-Hill

Name _____ Date _____

Meet Nancy Simmons

Read the Reading in Science pages in your book. Look for the main idea and details as you read. Remember, the main idea is the most important idea in the passage. Write the main idea in the chart below. Be sure to also write any details that help give more information about the main idea.

The Main Idea

Young bats grow and change to be like their parents.

Detail

Bat pups are small, pink, and have no hair when they are born.

Detail

Bat pups get milk from their mothers. It helps them grow bigger.

Detail

After a few months, the pups can fly, get their own food, and start their own families.

© Macmillan/McGraw-Hill

1. What did you learn about the false vampire bat? How did you learn it?

Possible answer: I learned that the false vampire bat is one of the largest bats in the world. I learned this from one of the photo captions.

2. What are baby bats called? What did you learn about how a young bat looks just after it is born?

Possible answer: I learned that baby bats are called pups. When a bat is born, it is small, pink, and has no hair.

Write About It

Find the Main Idea. How is a pup different from an adult bat? Use the chart you made to help you write your answer.

Possible answer: A bat pup looks different from an adult. It is small, pink, and has no hair. It cannot protect itself. Pups stay safe by holding on to their mothers. Adult bats have hair. They can find their own food and fly.

Staying Alive

Use your book to help you fill in the blanks.

Why do animals act and look the way they do?

1. Animals can _____change_____ , or adapt, to help them stay alive.

2. An _____adaptation_____ is a body part or a way an animal acts that helps it stay alive.

3. The long neck of a _____giraffe_____ is an adaptation.

4. The adaptation helps the giraffe _____eat_____ leaves from the tops of trees.

5. Some adaptations, like _____camouflage_____ , help animals hide from other animals.

6. Camouflage can be a color or a body _____shape_____ that helps an animal hide in nature.

7. A ptarmigan is a _____bird_____ that has brown feathers in the summer.

8. In the winter, the ptarmigan's feathers turn _____white_____ so it can blend in with the snow.

How do animals stay safe?

9. Some animals move in large ____groups____ to stay safe.

10. Staying together in a large group helps ____protect____ smaller fish from being eaten by bigger fish.

11. Other animals ____migrate____ to places where they can find food and stay warm during winter.

12. Some animals, like bears and mice, ____sleep____ during the cold winter.

Critical Thinking

13. What adaptations does a bear have to help it stay safe?

Possible answer: A bear has sharp claws and teeth

that help it catch food. Its dark fur helps it hide from

other animals.

_____

_____

_____

_____

Name _____ Date _____

Staying Alive

Describe each animal's adaptations to stay alive.

1.

giraffe

The giraffe has a long

neck so it can eat leaves

from tall trees and look

for other animals.

2.

stick bug

The stick bug looks just

like a leaf so it can stay

safe and fool animals

that want to eat it.

3.

zebra

The zebra has stripes to

confuse animals that hunt

it. Zebras stay in large

groups for protection.

4.

hawk

The hawk has wings so

it can fly. It has sharp

claws and a sharp beak

so it can hunt for food.

Staying Alive

Fill in the blanks. Use the words from the box.

adaptation	camouflage	groups	shape
blend	color	pattern	winter

There are many ways in which animals can stay safe. An _adaptation_ is a body part or a way an animal acts to stay alive. Giraffes have long necks to eat leaves from the tops of trees.

Some animals can _blend_ into their environment. The color or _shape_ of an animal can help it hide from other animals. This is called _camouflage_. The _pattern_ of spots on a leopard helps it hide. Some animals can grow fur and feathers of a different _color_. A ptarmigan has brown feathers in the summer, but in the _winter_ it will turn white. This helps it hide in the snow. Some animals travel in large _groups_. This prevents them from getting eaten.

Name _____ Date _____

Helpful Traits

Write About It

Describe an animal. Where does it live? What do you think it eats? What traits help it live in its environment?

Getting Ideas

Write the name of the animal you chose in the center circle. In the outer ovals, write details about the animal.

Possible answer:

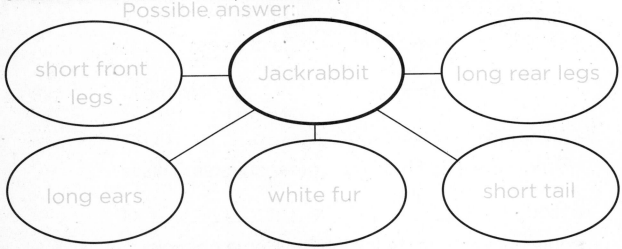

short front legs — Jackrabbit — long rear legs

long ears — white fur — short tail

Planning and Organizing

Clifton wrote three sentences about jackrabbits. Write Yes if the sentence describes them. Write No if it does not describe them.

1. _____Yes_____ They flatten their ears when they rest.

2. _____Yes_____ Some have white fur in the winter.

3. _____No_____ Jackrabbits have long tails.

Drafting

Write a sentence that tells what animal you are going to describe and where it lives.

Possible answer: Jackrabbits live in several different places.

Now write your description. Describe where it lives, what it eats, and what traits help it survive.

Revising and Proofreading

Fill in the blanks with descriptive words from the box.

flat	hind	short
front	long	

A jackrabbit has _____long_____ ears. Its _____front_____

legs are short, and its _____hind_____ legs are

longer. It also has a fairly _____short_____ tail.
Jackrabbits live just about everywhere in North

America. They live on _____flat_____ land and in

valleys. Some of them even live in the mountains.

Now revise and proofread your writing. Ask yourself:

▶ Did I describe this animal and its traits?

▶ Did I tell about traits that help it survive?

▶ Did I correct all mistakes?

Name _____ Date _____

Animals

Write the animal group next to each animal. Use the words in the box.

amphibian	fish	mammal
bird	insect	reptile

1.

finch

bird

4.

lizard

reptile

2.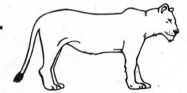

lion

mammal

5.

frog

amphibian

3.

bee

insect

6.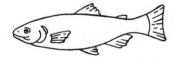

salmon

fish

Fill in the blanks. Use the words in the box.

adaptation	larva
camouflage	life cycle

I.

A ___life cycle___ shows how a living thing lives, grows, and dies.

2.

This beaver's teeth are an ___adaptation___ that help it live in its environment.

3.

A caterpillar is the ___larva___ of a butterfly.

4.

A toad uses ___camouflage___ to hide from other animals.

© Macmillan/McGraw-Hill

In Payment

By Aileen Fisher

Read the Unit Literature pages in your book.

 Write About It

Response to Literature

1. How does the butterfly "sort of pay for nibbles" in this poem?

The butterfly carries pollen to the different flowers.

_____

2. How will carrying pollen to the blossom help the butterfly?

More food will grow for the butterfly when it carries

pollen to the blossom.

3. What happens first, next, and last in this poem?

First, the caterpillar eats the leaves from plants. Next,

it makes a cocoon and turns into a butterfly. Last, the

grown butterfly carries pollen to different flowers.

_____

_____

Looking at Habitats

Fill in the important ideas as you read the chapter. Use the words in the box. You will use one of the words two times.

animals	lake	plants
food chains	nature	pond
forest	people	sea

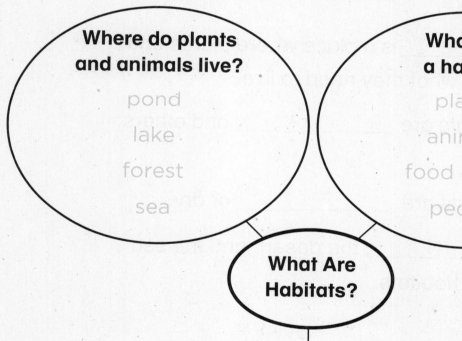

Where do plants and animals live?

pond

lake

forest

sea

What is in a habitat?

plants

animals

food chains

people

What Are Habitats?

Why do habitats change?

people

nature

drought

Name _____ Date _____

Places to Live

Use your book to help you fill in the blanks.

What is a habitat?

1. Animals need _____ food _____ , water, and shelter to live.

2. Plants need _____ soil _____ , water, and sunlight to live.

3. A _____ habitat _____ is a place where plants and animals find what they need to live.

4. Some habitats are _____ cold _____ and others are warm.

5. Other habitats are _____ rainy _____ or dry.

6. The _____ forest _____ , the desert, and the sea are kinds of habitats.

How do living things use their habitats?

7. Living things find _____ food _____ and shelter in their habitats.

8. Some animals eat the _____ plants _____ that grow in their habitats.

9. Some animals eat other _____ animals _____ that live in their habitats.

Critical Thinking

10. How do you think a snake survives in a very dry, sunny habitat?

   Possible answer: A snake can eat the animals that live

   in the same habitat. It can get water from the food it

   eats. It can hide underground to stay cool and safe.

   _____

   _____

   _____

   _____

Name _____ Date _____

Places to Live

Write how each living thing is using its habitat.

1.

fox

Possible answer: This fox has fur
that helps it stay warm in its habitat.
The fox uses its claws to dig a home
where it can stay safe in its habitat.

2.

cactus

Possible answer: This cactus has
thick stems and leaves in order to
store water. It has needles to stay
safe from animals that want to eat it.

3.

spider

Possible answer: This spider is
using its habitat to stay safe and
hunt for food.

Use with **Lesson 1**
Places to Live

© Macmillan/McGraw-Hill

Places to Live

Fill in the blanks. Use the words from the box.

habitat	shelter	tunnels
plants	sunlight	

Where can plants and animals live? Living

things can live in any ____habitat____ where they

get what they need to survive. Plants need soil,

nutrients, water, and ____sunlight____ from their

habitats in order to grow. Animals need food,

water, and ____shelter____ from their habitats in

order to grow.

Plants and animals use their habitats in

different ways. Some animals eat the

____plants____ and animals that live in their

habitats. Other animals dig ____tunnels____ in

the soil to hide from animals that want to eat them.

Some plants even eat animals that live in their

habitats!

Name _____ Date _____

Food Chains and Food Webs

Use your book to help you fill in the blanks.

What is a food chain?

1. A ___food chain___ shows how food energy moves from one living thing to another.

2. The ___Sun___ is at the beginning of all food chains.

3. Plants need sunlight in order to grow, and ___animals___ eat plants in order to live.

4. Some food chains involve animals that live in the ___water___, while others involve animals that live on land.

5. Some animals eat ___plants___ and animals that are no longer living.

6. Animals such as ___worms___ break up dead things into smaller pieces.

What is a food web?

7. A ____food web____ is two or more food chains that are connected.

8. A ____predator____ is an animal that hunts and eats other animals.

9. Animals that are hunted by other animals are called ____prey____ .

10. Sometimes, one kind of ____animal____ is food for many animals.

Critical Thinking

11. Describe a food chain that ends with a bird.

Possible answer: The Sun makes a tree grow. A beetle

eats the tree leaves. A bird eats the beetle.

_____

_____

_____

_____

© Macmillan/McGraw-Hill

Name _____ Date _____

Food Chains and Food Webs

These pictures show living things in a food chain.
Match each predator from the right column with its
prey in the left column.

Prey   **Predators**

1.

mouse

a.

owl

2.

moth

b.

brown bear

3.
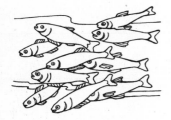

salmon

c.

kingsnake

4.

rattlesnake

d.

toad

Food Chains and Food Webs

Fill in the blanks. Use the words from the box.

break	food web	predator	study
food chain	plants	prey	Sun

Different living things need different kinds of

food in order to survive. A ___food chain___ shows

what an animal eats and where its food comes

from. Scientists ___study___ food chains to

learn more about living things in our world.

All food chains start with the ___Sun___ .

Plants use light and heat from the Sun to grow,

then animals eat the plants. A ___predator___

is an animal that eats other animals. An animal

that is hunted by a predator is called ___prey___ .

Some living things eat nonliving ___plants___ and

animals. They ___break___ down the dead

parts into pieces that become part of the soil. One

kind of animal can be food for many animals. A

___food web___ shows how different food chains

are connected. You are part of a food web too!

Name _____ Date _____

A Food Web for Lunch

✏ Write About It

Explain how Emma, the chicken, the lettuce, and the wheat form a food web. Think about the food chains in Emma's lunch to help you form a food web of your own lunch.

Getting Ideas

Create a food web for your lunch.

Planning and Organizing

Put the steps in the correct order.

_____3_____ Emma drinks milk for breakfast.

_____1_____ The cow eats grass.

_____2_____ A farmer milks the cow.

Drafting

Write a sentence to explain the food web. Tell your main idea.

Possible answer: Emma's breakfast is made up of different food chains.

Now write how the foods in Emma's breakfast form a food web. Start with the sentence you wrote above. Explain how the foods are connected.

Revising and Proofreading

Zack wrote some sentences. He made five mistakes. Find the mistakes. Then correct them.

The Son is the most important part of the
food web. It gives energie to plants. The plants
is eaten by the animals. Some animals then
produce food. Chickens lay eggs. cows produce
milk. Farmers gather the eggs for people to eat.
Farmers also milk cows and bottle the milk.
People drink the milk

Now revise and proofread your writing. Ask yourself:

▶ Did I explain the food web in Emma's breakfast?

▶ Did I tell the steps in order?

▶ Did I correct all mistakes?

Name _____ Date _____

Habitats Change

Use your book to help you fill in the blanks.

How do habitats change?

1. Habitats _____change_____ in many different ways.

2. Nature can make habitats change _____slowly_____ or quickly.

3. A drought is a slow change that takes place when an area gets little or no _____rain_____ for a long time.

4. Animals and _____people_____ can change habitats.

What happens when habitats change?

5. When habitats change, the _____plants_____ and animals that live there may adapt or make changes.

6. Other plants and animals may not be able to _____survive_____ and can become endangered.

7. An animal becomes _____endangered_____ when many of its same kind die.

© Macmillan/McGraw-Hill

How can we tell what a habitat used to be like?

8. Scientists study _____fossils_____ to learn what Earth was like long ago.

9. Fossils can tell scientists how _____habitats_____, plants, and animals have changed over time.

10. Some fossils do not _____match_____ the habitat where they were found.

11. That tells scientists that there has been a _____change_____ in the habitat.

12. When an animal becomes _____extinct_____, there are no more of its kind left in the world.

Critical Thinking

13. Scientists have found fossils with fins and tails in dry areas. What do you think these places might have looked like long ago? How did they change?

Possible answer: Animals with fins and tails live in water. The places where these fossils were found probably had water long ago. Then, the water dried up, and the animals became extinct.

Name _____ Date _____

Habitats Change

**Use the picture to answer the questions. Use the
words in the box in your sentences.**

| drought | endangered | extinct | fossil |

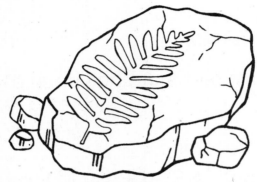

1. How do you think this habitat has changed
over time?

Possible answer: This plant fossil looks very different

from the desert plants. This means that long ago, this

habitat had different kinds of plants, and this plant is

now extinct.

2. How do you think this habitat became a desert?

Possible answer: I think that there was a drought in this

habitat. There was little or no rain for a long time and

the land slowly dried up.

Habitats Change

Fill in the blanks. Use the words from the box.

change	endangered	fossil	people
drought	extinct	habitat	

Plants and animals live in different places. A

_____habitat_____ is a place where plants and

animals live. People also live in habitats. Habitats

can _____change_____ over time. A _____drought_____

changes a habitat when an area gets little or no

rain for a long time. Habitats can change because

of _____people_____ , too. People destroy plant

and animal homes by building roads and buildings.

When habitats change, plants and animals may

die. A plant or animal becomes _____endangered_____

when there are only a few of its kind left in the world.

A plant or animal becomes _____extinct_____

when there are no more of its kind left. When plants

or animals disappear, they may leave a _____fossil_____

behind. Scientists study fossils to learn what Earth

was like long ago.

© Macmillan/McGraw-Hill

Name _____ Date _____

Meet Mike Novacek

Read the Reading in Science pages in your book. As you read, think about how Mike and his team classify and categorize the fossils they discover. Mike has collected fossils of reptiles, mammals, and dinosaurs.

Use the chart below to classify the animals you have learned about. Remember, when you classify and categorize, you compare things. Then you put the ones that are alike into groups.

Fossils

Reptile	Mammal	Dinosaur
Possible response: alligator, salamander, snake	Possible response: saber-toothed cat, mammoth, Kryptobaatar	Possible response: Ankylosaur, Lambeosaurus, Velociraptor

1. Where did you put the fossil of the Kryptobaatar skull in the chart?

I put it in the mammal column.

© Macmillan/McGraw-Hill

✏ Write About It

1. Classify and categorize. How can you put fossils into groups?

I can group fossils by looking at things that are the same, such as animals with backbones or animals that only ate plants.

2. Why do you think scientists travel around the world looking for fossils?

Possible answer: Scientists look for fossils around the world to compare animal groups.

3. What do you think a Kryptobaatar looked like? Draw a picture.

Drawings will vary.

Name _____ Date _____

Looking at Habitats

Fill in the blanks. Use the words in the box.

drought	extinct	predator
endangered	food chain	prey

1. An animal that hunts and eats another animal is

    called a ____predator____ .

2. An animal that is eaten by another animal is

    called ____prey____ .

3. A ____food chain____ shows what an animal eats
    and where it gets its food.

4. An animal becomes ____endangered____ when
    there are only a few of its kind left on Earth.

5. When an animal becomes ____extinct____ ,
    there are no more of its kind living on Earth.

6. A ____drought____ happens when a place gets
    little or no rain for a long time.

Draw pictures to complete the food chain.

hawk

Possible drawings may include prairie lizards and field mice or other small rodents.

grasshopper

Possible drawings may include grasses and other low-lying plants.

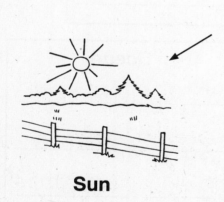

Sun

I. What is at the beginning of this food chain?

The Sun is at the beginning of the food chain.

2. Is the animal that comes after the grasshopper a kind of predator or a kind of prey? Explain.

The animal that comes after the grasshopper is a kind

of predator because it eats the grasshopper. Animals

that eat other animals are predators.

Name _____ Date _____

Kinds of Habitats

Fill in the important ideas as you read the chapter.

Woodland Forest
warm in summer, cool
in winter, leaves turn
color, some animals
sleep all winter

Tropical Forest
warm, steamy
climate, tall trees
with big leaves,
animals live in trees

Pond
small body of fresh
water, doesn't flow,
plants grow in shallow
water near shore

Kinds of Habitats

Ocean
large body of salt
water, always
flowing, covers most
of Earth

Arctic
cool desert near the
North Pole, no trees,
plants grow low to
the ground, windy

Desert
dry, very little rain,
hot during the day,
cool at night, sandy
and rocky soil

© Macmillan/McGraw-Hill

Forests

Use your book to help you fill in the blanks.

What is a woodland forest like?

1. A ___woodland forest___ habitat has many trees.

2. It is warm in the summer and ___cool___ in the winter.

3. A habitat is a place where ___plants___ and animals get what they need to live.

4. Most ___trees___ in the forest have leaves that change color in the fall.

5. Some trees have leaves that stay ___green___ all year.

6. Animals can ___survive___ in a woodland forest in many ways.

7. Some animals eat leaves, ___fruit___, and nuts.

8. Other animals build homes in trees and ___sleep___ in logs during the winter.

Name _____ Date _____

What is a tropical rain forest?

9. A ___tropical___ rain forest is a warm, steamy, moist place with many trees.

10. Some animals, such as birds, bats, and insects, live high in the ___treetops___ .

11. Other animals such as jaguars, tapirs, and wild boars live on the ___ground___ .

12. Many trees grow very tall, have large ___leaves___ , and block sunlight from falling to the ground below.

Critical Thinking

13. Why do you think animals in the tropical rain forest do not sleep all winter?

Possible answer: The tropical rain forest is warm,

steamy, and moist. It does not get cold, so animals

do not have to sleep all winter. There is plenty of

food all year.

_____

© Macmillan/McGraw-Hill

Forests

How do woodland forests and tropical rain forests compare? Fill in the Venn diagram.

Woodland Forest   **Both**   **Tropical Rain Forest**

cool in the winter; warm in the summer; some leaves change color and fall off

plants and animals live there; tall trees can block sunlight

warm, moist, and steamy; has leaves that stay green all year

Forests

Fill in the blanks. Use the words from the box.

animals	rain forest	survive	woodland
color	sunlight	winter	

A habitat is a place where plants and animals

get what they need to live. A ____woodland____

forest is one kind of habitat. It has many trees. It is

cold during ____winter____ and warm during

summer. Many of the trees have leaves that

change ____color____ and drop to the ground

in the fall. Plants and animals ____survive____ in

this kind of forest in many ways. Some animals

use the trees as their homes. Others sleep during

the winter to survive.

A tropical ____rain forest____ is warm, steamy,

and moist. The trees are tall and have very large

leaves. They block ____sunlight____ from getting

to the ground. Some ____animals____ live in the

treetops. Other animals live on the ground.

Name _____ Date _____

Meet Liliana Dávalos

Read the Reading in Science pages in your book.
As you read, think about how Liliana compares and
contrasts things in her work as a biologist at the
American Museum of Natural History. Remember,
when you compare things, you decide how they are
alike. To contrast is to decide how things are different.

Answer the questions and fill in the chart below.

1. What other habitats have you learned about in
this chapter?

I learned about forests, deserts, ponds, and oceans.

2. How is the rain forest alike and different from
other kinds of forests?

Rain Forest	Regular Forest	Both
some in South America; gets a lot of rain; home to many animals	in other parts of the world; doesn't get as much rain; home to different animals	people can cause changes to these habitats and endanger animals

Name _____ Date _____

✏ Write About It

1. Compare and Contrast. How would life change for the manakins if the Amazon rain forest were cut down? Would it be the same as it is today? Explain.

The manakins might become extinct if the rain forest were cut down. This is because if the rain forest were gone, they would have no habitat where they could get what they need to live.

2. A biologist is a scientist who studies living creatures. What other kinds of scientists have you learned about? How are they alike and different?

I have learned about geologists, paleontologists, and archaeologists. They study different things, but they all try to learn more about Earth and our world.

3. Biologists, like Liliana, often compare and contrast animals in their work. Why?

Biologists compare and contrast living things to learn more about how they look and where they live.

Hot and Cold Deserts

Use your book to help you fill in the blanks.

What is a hot desert like?

1. A _____desert_____ is a very dry and sandy habitat.

2. This kind of habitat can be _____hot_____ during the day and cool at night.

3. It does not _____rain_____ often in the desert.

4. Plants in this habitat survive by storing _____water_____ in their stems and leaves.

5. Some desert plants have _____roots_____ that spread far out from the plant.

6. Desert animals get water from eating _____plants_____ or other animals.

7. Most desert animals sleep during the day and hunt for _____food_____ at night.

What is the Arctic like?

8. The _____Arctic_____ is a very cold and windy desert near the North Pole.

9. Many animals that live in this habitat have thick

   _____fur_____ that keeps them warm.

10. Other animals have a thick layer of fat, called

   _____blubber_____ , to keep warm.

11. Small, low plants grow in the Arctic to stay safe

   from the cold _____winds_____ .

Critical Thinking

12. Do you think that plants in hot and cold desert habitats store water in the same way? Why or why not?

   Possible answer: No, I don't think that they do. Arctic

   plants have shallow roots because the ground is

   frozen. Desert plants store water in large leaves and

   deep roots.

   _____

   _____

© Macmillan/McGraw-Hill

Hot and Cold Deserts

If the sentence describes a hot desert, write Desert. If the sentence describes the Arctic, write Arctic.

1. It can be hot during the day, and cool at night.

   Desert

2. It is very windy and cold.

   Arctic

3. The plants store water in their stems.

   Desert

4. There are no trees.

   Arctic

5. Animals have thick blubber or fur.

   Arctic

6. Animals have light fur, feathers, or scales.

   Desert

7. Animals sleep during the day and hunt at night.

   Desert

8. It is near the North Pole.

   Arctic

Name _____ Date _____

Hot and Cold Deserts

Fill in the blanks. Use the words from the box.

Arctic	cacti	hunt
blubber	desert	North Pole

Some places on Earth get very little rain. These places are called deserts. A hot _____desert_____ can be very warm during the day and cool at night.

Some hot-desert plants, like _____cacti_____, store water in their thick stems. Many hot-desert animals sleep during the day and _____hunt_____ at night.

The _____Arctic_____ is a cold and windy desert near the _____North Pole_____. There are no trees, and plants grow low to the ground to stay safe from wind. Many cold-desert animals, like seals, have thick fur or _____blubber_____ to stay warm. Desert animals and plants have adaptations that help them survive in their habitat.

Oceans and Ponds

Use your book to help you fill in the blanks.

What is the ocean like?

1. The largest bodies of water on Earth are called
   ____oceans____ .

2. An ocean is a large body of ____salt____
   water.

3. Most of ____Earth____ is covered by oceans.

4. Kelp is a kind of ____seaweed____ , or ocean
   plant.

5. It grows in the ocean and provides ____food____
   for many ocean animals.

6. Animals in the ocean have ____body____
   parts that help them swim through the water.

7. Some animals in the ocean have ____shells____ ,
   spines, or stingers to help them stay safe.

Name _____ Date _____

What is a pond like?

8. A _____ is much smaller than an
   ocean.
   pond

9. Ponds have _____ water and do not
   flow.
   fresh

10. Frogs, fish, and _____ are some
    animals that live in or near ponds.
    turtles

11. Many plants grow in _____ pond
    water near the shore.
    shallow

12. Animals that live in ponds _____ in
    different ways.
    breathe

Critical Thinking

13. Do you think that the same types of animals
    live in both oceans and ponds?

    Possible answer: No, I don't. Ocean water is salty, and

    pond water is fresh. I think animals in each body of

    water have adaptations to help them survive in that

    environment.

    _____

    _____

Oceans and Ponds

Look at the animal and plant pictures beneath the box. Write the name of each animal or plant under the habitat where they live.

salamander	dolphin
cat tails	penguin
mosquito	coral reef

salamander

mosquito

dolphin

cat tails

penguin

coral reef

© Macmillan/McGraw-Hill

Oceans and Ponds

Fill in the blanks. Use the words from the box.

coral	fresh	kelp	pond
deep	habitat	ocean	shallow

Most of Earth is covered by water. An ____ocean____ is a large body of water that flows. Plants such as ____kelp____ grow in the water and provide food for animals. A special animal called ____coral____ lives on the ocean floor and provides shelter for many other animals. Some animals, such as mussels and crabs, live near the shore. Other animals, such as sea cucumbers and sea spiders, live in ____deep____ waters.

A ____pond____ is a body of water that does not flow. Most ponds have ____fresh____ water in them. Different kinds of plants and animals live in this ____habitat____ . Some plants grow in ____shallow____ water near the shore. Their stems and leaves rise to the top of the water.

© Macmillan/McGraw-Hill

A Visit to the Ocean

Write a story about a trip you might take to the ocean. How would you get there? Who would you go with? Describe in your story what you would see, hear, and do. Write how it might feel to be there.

Getting Ideas

Picture yourself standing on a beach next to the ocean. Write what you see and hear.

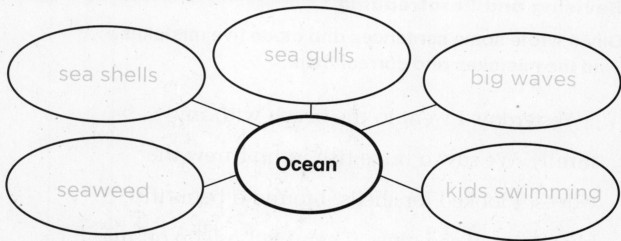

Planning and Organizing

Jackson wrote three sentences. They describe what he saw, heard, and did at the ocean. Circle the descriptive words he used.

1. The (gigantic) ocean waves roared (loudly).

2. I saw (white) gulls sitting on a (big) rock near the shore.

3. I found a piece of (green) sea glass and two (pretty) (pink) shells.

Drafting

Write a sentence to begin your story. Use I to tell about yourself. Tell where you went and when.

Possible answer: Yesterday, I went to the beach with my family.

Now write a story on a separate piece of paper. Put the events in time order. Describe what you saw, heard, and did at the ocean.

Revising and Proofreading

Olivia wrote some sentences and made five mistakes. Find the mistakes and correct them.

Yesterday, I went to the beech with my
family. We saw a huge fish jump threw the
waves. I looked for shells. I found a beautiful
blue peice of sea glass. Then I fell asleap on my
beach towel. When I wake up, it was almost
time to go home.

Now revise and proofread your writing. Ask yourself:

► Did I tell how I got to the ocean and with whom I went?

► Did I describe what I saw, heard, and did?

► Did I correct all mistakes?

Kinds of Habitats

Fill in the blanks. Use the words in the box.

Arctic	habitat	pond
desert	ocean	tropical

1. A ____desert____ is a place that gets very little rain.

2. A place where plants and animals live is called a ____habitat____ .

3. An ____ocean____ is a large body of water that flows.

4. A ____tropical____ rain forest is a place with many trees that is warm, steamy, and moist.

5. The cold desert near the North Pole is called the ____Arctic____ .

6. A small body of fresh water that does not flow is called a ____pond____ .

© Macmillan/McGraw-Hill

Name _____ Date _____

Identify each habitat.

1.

desert
_____

2.

pond
_____

3.

woodland forest
_____

4.

Arctic
_____

5.

ocean
_____

Earthworms

Soil Helpers

Read the Unit Literature pages in your book.

 Write About It

Response to Literature

1. What do you think would happen to soil if there were no earthworms?

I think there would be no new soil to mix, and roots

would not grow and spread so easily. Plants may not

get what they need to live.

2. Can you imagine what the world looks like to an earthworm? Use the article to give you ideas. Draw a picture.

Drawings will vary.

Name _____ Date _____

Land and Water

Fill in the important ideas as you read the chapter.
Use the words in the box.

continent	lake	ocean	stream
earthquake	landslide	plain	valley
flood	mountain	pond	volcano

What do you know about the Earth's land and water?

What is land like on Earth?

valley

continent

plain

mountain

What is water like on Earth?

ocean

lake

pond

stream

How can Earth's land and water change?

flood

landslide

volcano

earthquake

Earth's Resources

Fill in the important ideas as you read the chapter. Write at least one way we use each of the natural resources shown on the left. Then, answer the question.

How do we use Earth's resources?

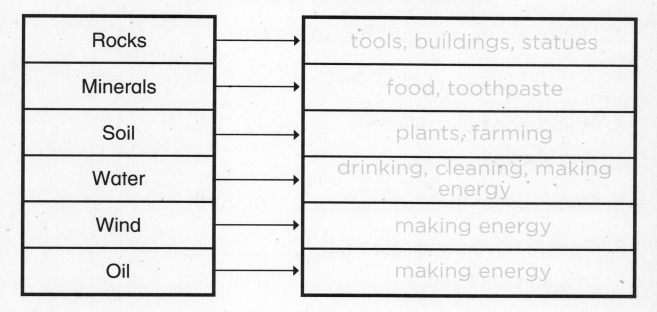

Rocks	tools, buildings, statues
Minerals	food, toothpaste
Soil	plants, farming
Water	drinking, cleaning, making energy
Wind	making energy
Oil	making energy

Why should we care for Earth's resources?

Possible answer: We should care for Earth's resources

because we need the resources in order to live. If we

pollute them or use them up, plants and animals may be

harmed and we may not be able to use them anymore.

Name _____ Date _____

Rocks and Minerals

Use your book to help you fill in the blanks.

What are rocks?

1. We use _____ like plants,
   animals, water, and rocks every day.

2. Unlike plants and animals, rocks are
   _____ resources.

3. Rocks can have different _____
   and shapes.

4. Rocks cover the _____ of Earth.

5. People have used rocks as _____ for
   thousands of years.

6. People can also use rocks to carve _____
   or build things.

What are minerals?

7. Most ____rocks____ are made of one or more minerals.

8. A ____mineral____ is a nonliving thing that comes from Earth.

9. It takes ____millions____ of years for rocks and minerals to form inside Earth.

10. People must ____dig____ to find rocks and minerals.

11. People use minerals like ____fluorite____ to help make toothpaste, steel, and other materials.

Critical Thinking

12. Why are rocks and minerals natural resources?

____Rocks and minerals are natural resources because____

____they come from Earth and people use them every____

____day. Walls and streets are made of rock. Pencils,____

____magnets, and table salt are made from minerals.____

_____

Name _____ Date _____

Rocks and Minerals

Fill in the blanks. Then find the vocabulary words in the puzzle.

1. A _____rock_____ is a hard, nonliving part of Earth.

2. A rock can be made of one _____mineral_____ or made of many different kinds.

3. A _____natural_____ resource is something from nature that people use.

4. People can make _____tools_____ out of rocks.

5. The mineral _____graphite_____ can be found in a pencil.

N	A	T	U	R	A	L	S	T	T
R	M	I	N	E	R	A	L	F	O
O	K	P	M	T	R	B	I	U	O
C	F	G	A	E	T	O	M	S	L
K	G	R	A	P	H	I	T	E	S

© Macmillan/McGraw-Hill

Name _____ Date _____

Rocks and Minerals

Fill in the blanks. Use the words from the box.

graphite	natural resource	tools
magnetite	statues	
minerals	surface	

Rocks are the most common materials on Earth.

They cover the _____surface_____ of Earth, from the

top of a mountain to the bottom of the ocean. Rocks

and _____minerals_____ are nonliving things that make

up part of Earth's surface.

Rocks and minerals are natural resources. A

_____natural resource_____ is something from nature, such

as water, wood, or minerals, that people use in

everyday life. The mineral _____magnetite_____ is found

in magnets, and _____graphite_____ is found in pencils.

For thousands of years, people have made _____tools_____

from rocks. They have even made _____statues_____

from rocks. The Sphinx in Egypt was carved from

rock thousands of years ago.

© Macmillan/McGraw-Hill

Name _____ Date _____

Rock and Stroll

✏ **Write About It**

Write a letter to a friend. Write about a walk you took. Describe the rocks you saw. Explain how you think they got their shape.

Getting Ideas

Fill in the chart. In the first column, tell what rocks you saw. In the second column, describe them.

Types of Rocks	Details
boulder	black, pieces of mica
stones bordering path	smooth, oval, brown and gray
pebbles in path	round, small, pink and gray
marble in statue	smooth, shiny, gray, streaks of red

Planning and Organizing

Write Yes if the sentence describes a rock. Write No if it does not.

1. _____Yes._____ The big boulder was gray and black.

2. _____Yes_____ The small stones were smooth and oval.

3. _____No_____ I really like to climb rocks at the beach.

Drafting

Write a greeting and first sentence for your letter. It should tell where you took your walk.

Dear Lucy,
_____

_____ This morning, Zoe and I walked in the park. _____

Now write your letter on a separate piece of paper. Describe the rocks you saw, and sign your name.

Revising and Proofreading

Fill in the blanks with words from the box.

gigantic	heavy	tall
gray	small	weird

This morning, Zoe and I walked in the park. We saw

a _____ gigantic _____ rock. It was very _____ tall _____ .

The stone was a deep _____ gray _____ . The rock had

a _____ weird _____ shape. I think that the _____ heavy _____

rain wore the rock down. The rocks by a pond were

very _____ small _____ and white.

Now revise and proofread your writing. Ask yourself:

▶ Did I describe the rocks and how they got their shapes?

▶ Did I correct all mistakes?

Name _____ Date _____

Soil

Use your book to help you fill in the blanks.

What is soil?

1. Earth's _____soil_____ is made of a mix of sand, clay, rocks, and minerals.

2. Parts of _____plants_____ and animals that have died are in soil, too.

3. Clay soil, topsoil, and _____sandy soil_____ are found in different places and have different colors.

4. Each kind of soil feels different and has a different _____texture_____ .

5. Some soils feel like _____clay_____ or pebbles.

6. Other soils feel _____sandy_____ and are light in color.

7. Some soils hold more _____water_____ than others.

8. The soils that hold more water have a _____darker_____ color.

© Macmillan/McGraw-Hill

How is soil formed?

9. It can take a very long time for rocks and

   _____minerals_____ to break down into soil.

10. When plants and animals die, their parts

    _____decompose_____ and rot away.

11. The _____nutrients_____ that were once inside

    living things make the soil healthy for plants.

12. Plants grow best in _____topsoil_____.

13. Topsoil is the _____layer_____ of soil with

    decaying plant and animal parts.

14. A mix of soil and parts of rotting plants and

    animals is called a _____compost_____ pile.

Critical Thinking

15. Why is soil important?

    Soil is important because we use it to grow plants.

    Soil can hold water to help plants grow. Once plants

    grow, we can eat the plants and feed them to animals.

Name _____ Date _____

Soil

Match each word in the box to the correct picture and use the word in a sentence.

compost	decompose	topsoil

1.

Possible answer: Parts of dead plants and animals decompose and rot into the soil.

2.

Possible answer: Compost is a mix of soil and rotting plant parts that can help make soil healthy.

3.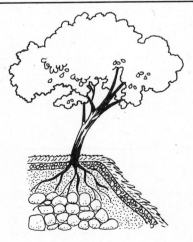

Possible answer: Plants grow best in topsoil.

Name _____ Date _____

Soil

Fill in the blanks. Use the words from the box.

decompose	natural resources	rocks
layer	nutrients	texture

Soil can be found almost everywhere on land.

Soil is one of Earth's most important _natural resources_ .

Soil is formed when _rocks_ and minerals
break down into smaller pieces over many years.

Parts of dead plants and animals _decompose_

and become part of the soil, too. The _nutrients_
inside these once-living things help make the
soil healthy.

   Plants grow best in the top _layer_ of
soil, called topsoil. This is where the soil is richest
with nutrients. Some soils are light, and others are dark.

Each soil feels different and has a different _texture_ .
Some soils hold a lot of water, while others are sandy
and do not hold much water. However, all soils are
important to Earth.

Name _____ Date _____

Using Earth's Resources

Use your book to help you fill in the blanks.

How do we use natural resources?

1. People use air, wind, water, rocks, and soil as
   ___natural resources___ every day.

2. Earth can quickly ___replace___ resources
   such as water and wind.

3. Other resources, such as ___minerals___ , can
   not be made quickly by Earth.

Why should we care for Earth's resources?

4. It is important to care for Earth's ___land___ ,
   water, and air.

5. Pollution can harm living and ___nonliving___
   things such as plants, animals, and people.

6. Pollution makes Earth's air, ___water___ ,
   and land dirty.

7. To stop land pollution, people can clean up the
   ___litter___ they leave behind.

How can we save Earth's resources?

8. People can help to ____conserve____ Earth's resources.

9. Remember the ____3____ Rs: reduce, reuse, and recycle.

10. When people ____reduce____, they cut back on how much they use a resource.

11. When people ____reuse____ something, they use it again, often in a new way.

12. When people ____recycle____ glass, paper, and cans, they make new things out of them and reduce litter.

Critical Thinking

13. How do you use natural resources every day?

Possible answer: I breathe air and drink water every day. I eat food that comes from plants and animals. My paper and pencils are made from trees. My shirt is made of wool, which comes from sheep.

Name _____ Date _____

Using Earth's Resources

Each picture below shows a way to conserve Earth's natural resources. Write reduce, reuse, or recycle under the correct picture.

1.

reuse

2.

reuse

3.

reduce

4.

recycle

What are other ways you can help conserve Earth's resources where you live?

Possible answer: I can reduce trash by using the back of

my paper to draw. I can reduce the water I use by taking

showers. I can reuse my juice boxes to plant seeds.

© Macmillan/McGraw-Hill

Using Earth's Resources

Fill in the blanks. Use the words from the box.

conserve	pollution	wind
litter	recycle	
natural resources	reduce	

Earth needs your help. Every day, you use <u>natural resources</u> such as air, water, and land. Earth can quickly replace resources like water and <u>wind</u>. Resources such as minerals take longer to replace.

It is important to <u>conserve</u> Earth's resources.

Something that makes air, water, or land dirty is called <u>pollution</u>. Help keep land and water clean by picking up <u>litter</u>. You can protect resources if you <u>reduce</u> and reuse things. You can <u>recycle</u> paper, glass, and plastic so they can be made into something else. Remembering the 3 Rs is the first step to helping save Earth's resources.

Name _____ Date _____

A World of Wool

Read the Reading in Science pages in your book. As you read, pay attention to the most important ideas. List them in the chart below. Then summarize the article. Remember, when you summarize, you retell the most important ideas in the selection.

Idea #1
Scientists learn how people in other countries get wool.

Idea #2
In Peru, Juana wears llama wool sweaters to stay warm.

Idea #3
Llamas have thick fur. Some farmers raise llamas for their fur.

Summary
People get wool from different animals. Juana lives in a cold place. She wears sweaters made from llama wool. Llamas grow long fur to stay warm. Some farmers raise llamas for their fur.

© Macmillan/McGraw-Hill

🖊 **Write About It**

Summarize. Write a paragraph that retells what you learned about llama wool. Use the following words in your writing: cold, warm, sweaters, llamas, camels, fur, spin, yarn, clothes, Andes Mountains.

Llamas look like small camels. They live in cold places
like the Andes Mountains. Llamas grow thick fur to
help them stay warm in their habitat. Workers shave
the llamas and spin their fur into yarn. They can use
the yarn to make sweaters and other kinds of clothes.

_____

_____

_____

_____

_____

_____

Name _____ Date _____

Earth's Resources

Write a short story about what is happening in the picture. Use at least three words from the box.

conserve	natural resources	reduce
litter	pollution	reuse
minerals	recycle	

Title: Possible title: Cleaning the Park

Story: Possible story: These children from Mrs. Wood's

class are fighting pollution in the park. Cassie and Carrie are

planting a tree. This will help them save natural resources.

Alex is helping to recycle cans and reduce litter. These

children know that cleaning the park can help protect

Earth's resources.

© Macmillan/McGraw-Hill

If the sentence is true, write TRUE. If the sentence is not true, write FALSE.

1. ___TRUE___ Rocks are made of minerals.

2. ___TRUE___ Litter is garbage that people leave behind.

3. ___FALSE___ Plastic is a natural resource.

4. ___TRUE___ When dead plants or animals decompose, their parts rot away.

5. ___FALSE___ Soil is made only of rocks.

6. ___FALSE___ A compost is a mix of paper, plastic, and glass.

Name _____ Date _____

Sunflakes

By Frank Asch

Read the Unit Literature pages in your book.

 **Write About It**

Response to Literature

1. What season is the poet writing about? Use the poem to tell how you know.

   The poet is writing about summer. I know this because

   he talks about sleighing in July. July is a summer month.

2. What are some things that you do in July? How do your activities compare to the poet's?

   I go to the beach. I go to summer camp. The poet

   talks about going sleighing and having sunflake fights.

3. What do you think a sunflake looks like? Draw a picture.

Observing Weather

Fill in the important ideas as you read the chapter.

Weather Words

precipitation

temperature

Weather Tools

thermometer

anemometer

wind sock

rain gauge

weather vane

**How Can
We Describe
Weather?**

Weather Changes

clouds

storms

hurricanes

tornadoes

Name _____ Date _____

Weather

Use your book to help you fill in the blanks.

What is weather?

1. People think about the _____weather_____ every day.

2. The _____temperature_____ outside helps people choose what kind of clothes to wear.

3. Temperature is a measure of how _____hot_____ or cold something is.

4. People use a _____thermometer_____ to measure temperature.

5. There are _____two_____ ways to describe temperature: in degrees Fahrenheit or degrees Celsius.

6. The _____precipitation_____ that falls from the clouds can also be measured.

7. Rain, snow, sleet, and _____hail_____ are kinds of precipitation.

What is wind?

8. The differences between hot and cold air

cause air to move, making _____wind_____ .

9. You can use a _____wind sock_____ to measure the
direction of wind.

10. This tool also shows how _____strong_____ the
wind is blowing.

11. People can use an _____anemometer_____ to measure
the speed of the wind.

Critical Thinking

12. What is wind? What can wind tell you about
weather?

Possible answer: Wind is moving air. It is caused by

differences in hot and cold air. The wind can tell how

the air is moving outside, and whether it is hot or cold.

_____

_____

_____

Name _____ Date _____

Weather

Draw a line to match the weather tool with what it measures.

1.

        a. temperature

2.

        b. wind speed

3.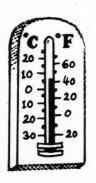

        c. precipitation

Weather

Fill in the blanks. Use the words from the box.

anemometer	rain gauge	weather
Fahrenheit	temperature	wind
precipitation	thermometer	wind sock

Look out the window. What is the ___weather___ like? Is it sunny? Is it rainy? People use special tools to find out about the weather. A ___thermometer___ is used to find out how hot or cold it is outside.

This tool measures the ___temperature___ of the air. The air is measured in degrees ___Fahrenheit___ or in degrees Celsius.

Moving air is called ___wind___. The speed with which the wind blows is measured by using an ___anemometer___. A ___wind sock___ shows what direction the wind is blowing. Rain, snow, sleet, and hail are kinds of ___precipitation___. A ___rain gauge___ is used to measure precipitation. These tools help people learn about the weather.

Name _____ Date _____

A Snowy Day

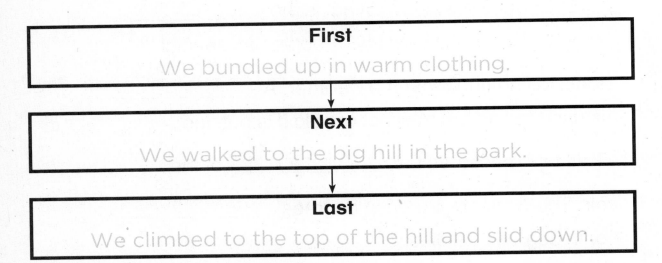 **Write About It**

Write a story about what you might do on a snowy day.

Getting Ideas

Picture a snowy day in your mind. Now put yourself in the picture. Write what you are doing.

First
We bundled up in warm clothing.

↓

Next
We walked to the big hill in the park.

↓

Last
We climbed to the top of the hill and slid down.

Planning and Organizing

Put the sentences in time order.

_____1_____ We bundled up in warm clothing.

_____3_____ We climbed to the top of the hill and slid down.

_____2_____ We walked to the big hill in the park.

Drafting

Write the first sentence of your story. Tell how you started your snowy day.

Possible answer: I was so excited when I saw that it was
_____
snowing.

Now write your story on a separate piece of paper. Put the events in time order. Include details.

Revising and Proofreading

Use the words in the box to fill in the blanks.

cold	long	warm
huge	soft	

It was a cloudy and _____cold_____ day. Andy and

I wore _____warm_____ clothes outside. We noticed

_____long_____ , narrow icicles hanging from the

trees. They were beautiful! Maple Hill was covered

in _____soft_____ , deep snow that made it hard

to climb. At the top, we made a _____huge_____ ball

of snow. Then we rolled it down the hill.

Now revise and proofread your writing. Ask yourself:

▶ Did I use details to tell what I might do on a snowy day?

▶ Did I correct all mistakes?

Name _____ Date _____

The Water Cycle

Use your book to help you fill in the blanks.

How does water disappear?

1. Water ___evaporates___ when it gets very warm.

2. When water evaporates, it changes from a ___liquid___ to water vapor.

3. Water vapor is in the form of a ___gas___ .

4. When water ___condenses___ , it changes from a gas to a liquid.

5. When the air ___cools___ , the water vapor turns back into tiny droplets of water.

6. These droplets can form ___clouds___ in the sky.

What is the water cycle?

7. The _____water cycle_____ shows how Earth's water evaporates to form bodies of water, and then condenses.

8. When water is warmed by the _____Sun_____, it evaporates.

9. The air forms _____clouds_____ when the water vapor in the air condenses.

10. Rain and _____snow_____ then fall, and the water flows back to the oceans, rivers, and streams.

Critical Thinking

11. If there were no oceans, streams, rivers, or lakes, do you think it would still rain? Why or why not?

   Possible answer: I think it would not rain, because there would be no way for the water cycle to continue.

   _____

   _____

   _____

Name _____ Date _____

The Water Cycle

Describe what happens in each step of the water cycle.

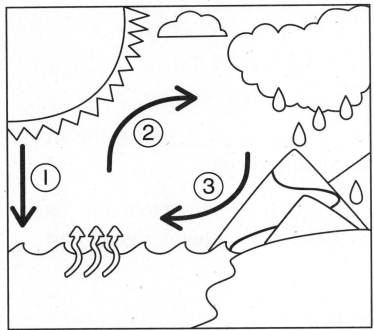

In the first step, the Sun heats the water until it

evaporates. In the second step, the air cools and the

water vapor condenses and forms clouds. In the third

step the rain or snow falls from the clouds to Earth, and

that water flows back into the oceans.

...erving Weather

The Water Cycle

Fill in the blanks. Use the words from the box.

clouds	flows	water cycle
condenses	rain	water vapor
evaporates	Sun	

How does water change? The ___water cycle___
shows how water moves from Earth to the sky,

and back down again. The ___Sun___
warms the water in oceans, rivers, and streams.

The water ___evaporates___ , or turns into a gas

and rises. This gas is called ___water vapor___ .
When the air gets cooler, the water

___condenses___ , or turns back into a liquid. Tiny

droplets of water form ___clouds___ in the sky.

Precipitation like ___rain___ and snow

can fall from the clouds. The water ___flows___

down the land and into the oceans, rivers, and streams.
Then the cycle begins again.

Changes in Weather

Use your book to help you fill in the blanks.

What are different kinds of clouds?

1. Clouds can tell about changes in the _____ .

2. Small, puffy clouds that can appear in long

   rows are called _____ clouds.

3. During the _____ , it is easy to see

   these clouds.

4. Thin clouds that are very high in the sky are

   called _____ clouds.

5. Cirrus clouds are made of _____ .

6. Thick or thin clouds that are very low in the sky

   are called _____ clouds.

How can we stay safe from weather?

7. Weather changes when different kinds of

_____air_____ come together.

8. Storm clouds can gather, and ___lightning___ can form inside of them.

9. Very strong storms can cause ___disasters___ .

10. Thunderstorms with spinning columns of air are

called ___tornadoes___ .

Critical Thinking

11. How would you stay safe during a strong thunderstorm?

Possible answer: I would avoid open spaces outside. I would stay out of the water. I would not stand under a tree or any other tall object. I would go in a basement or another safe place indoors.

_____

_____

Name _____ Date _____

Changes in Weather

Use the words in the box to tell which clouds are
shown in the pictures.

cirrus	hurricane	tornado
cumulus	stratus	

1.

stratus

4.

hurricane

2.

tornado

5.

cumulus

3.

cirrus

Changes in Weather

Fill in the blanks. Use the words from the box.

cirrus	disasters	rows	tornado
cumulus	hurricane	stratus	weather

There are many different kinds of clouds. Clouds

tell about changes in the _____weather_____ .

Small, white, puffy clouds are called _____cumulus_____

clouds. They appear in long _____rows_____

and mean fair weather. Thin clouds that are very

high in the sky are called _____cirrus_____ clouds.

These clouds are made of ice. Thick or thin clouds

that cover the entire sky are called _____stratus_____

clouds. These clouds mean that rain or snow

is coming.

Weather can change when different types of air

come together. Very strong storms can cause

_____disasters_____ like floods. A _____hurricane_____ is

a storm with very strong winds. A _____tornado_____

is a column of spinning air. People can stay safe

from many storms by staying indoors.

Name _____ Date _____

Predicting Storms

Read the Reading in Science pages in your book. As you read, pay attention to the most important ideas. List them in the chart below. Then summarize the article. Remember, when you summarize, you retell the most important ideas in the selection.

Idea #1

Weather balloons travel above Earth. They collect data about wind and weather.

Idea #2

Doppler radar measures how fast a storm is moving and where it is going.

Summary

Scientists use different tools such as weather balloons and Doppler radar to predict the weather. Their predictions help many people.

Write About It

Summarize. How does Doppler radar work?

Doppler radar works by using radio waves to measure

how fast a storm is moving and in which direction it

is moving.

Write a paragraph that retells what you learned about why scientists try to predict the weather.

Weather can change very quickly, and sudden storms can

be very dangerous. Scientists try to predict the weather

to help people stay safe during a storm.

Observing Weather

Fill in the blanks. Use the words in the box.

condenses	precipitation	water cycle
evaporates	temperature	water vapor

1. The ___water cycle___ shows how water changes on Earth.

2. When water ___evaporates___, it changes from a liquid to a gas.

3. When water ___condenses___, it changes from a gas to a liquid.

4. To find out hot or cold something is, we can measure its ___temperature___.

5. Rain, snow, sleet, and hail are all different kinds of ___precipitation___.

6. When water is a gas, it is called ___water vapor___.

Solve each riddle.

1. I am thin and high in the sky. I am made of ice. What kind of cloud am I?

   cirrus

2. I am small, white, and puffy. I appear when the weather is fair. What kind of cloud am I?

   cumulus

3. I am low in the sky. I appear when rain or snow is on the way. What kind of cloud am I?

   stratus

4. I am a tool that can measure the speed of the wind. What am I?

   anemometer

5. I am a spinning column of air. I can cause a lot of damage. What am I?

   tornado

Name _____ Date _____

Earth and Space

Fill in the important ideas as you read the chapter.
Use the words in the box.

axis	orbit	planet	solar system
Moon	phase	rotation	

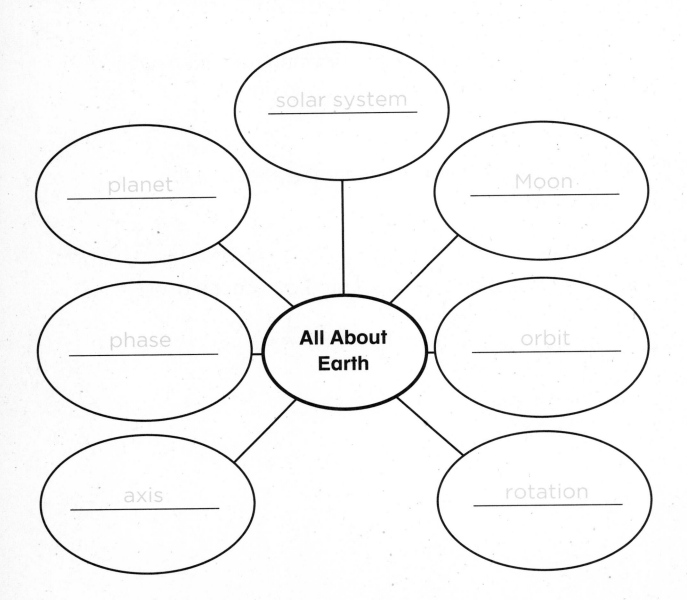

solar system

planet

Moon

phase

All About Earth

orbit

axis

rotation

Day and Night

Use your book to help you fill in the blanks.

What causes day and night?

1. Earth's ____rotation____ is what causes day and night.

2. It is ____day____ when our side of Earth faces the Sun.

3. When our side of Earth faces the Sun, it is ____night____ on the other side.

4. Earth always ____rotates____ in the same direction.

5. It takes 24 hours for Earth to make one full turn on its ____axis____ .

6. The axis is an imaginary line that goes through the ____middle____ of Earth.

Name _____ Date _____

Why do the Sun and Moon seem to move?

7. The _____Sun_____ seems to move across the sky during the day.

8. Shadows on the ground change as Earth _____rotates_____ .

9. At night, the _____Moon_____ seems to move, too.

10. This is because _____Earth_____ is rotating.

Critical Thinking

11. What happens on the other side of Earth when it is night where you live? How do you know?

When it is night where I live, it is day on the other side

of Earth. I know because Earth is round. When it is

dark on one side of Earth, the other side is facing the

Sun.

_____

_____

_____

_____

Day and Night

Fill in the blanks. Use the words from the box.

| axis | Day | night | rotation |

1. _____Day_____ and night are caused by Earth's rotation.

2. Earth's ____rotation____ never changes direction.

3. Every 24 hours, Earth rotates once on its ____axis____ .

4. When it is day where you live, it is ____night____ on the other side of the world.

Name _____ Date _____

Day and Night

Fill in the blanks. Use the words from the box.

axis	Earth	rotation	Sun
day	night	shadows	

You cannot feel it, but you are spinning right

now. In fact, _____Earth_____ is always spinning.

It spins all _____day_____ and all night. It even

spins when you are asleep! This turning is called

Earth's _____rotation_____ . It is why we have day

and _____night_____ .

Every 24 hours, Earth rotates one time on its

_____axis_____ . As it rotates, light from the

_____Sun_____ lights a different part of the

planet. This is why _____shadows_____ are longer

during the day. When it is day on one side of the

world, it is night on the other side.

© Macmillan/McGraw-Hill

Why Seasons Happen

Use your book to help you fill in the blanks.

What are the seasons like?

1. In the fall, the _____ weather _____ is cool.

2. Some leaves _____ turn _____ colors and fall off their trees.

3. The air is much colder during the _____ winter _____ .

4. In some places, the cold rain turns to _____ snow _____ .

5. Some animals, like birds, _____ fly _____ to warmer places.

6. People wear warmer _____ clothing _____ .

7. In the spring, _____ rainy _____ days help new plants grow.

8. Summer is the warmest _____ season _____ of all!

What causes the seasons?

9. Earth takes about 365 days to _____orbit_____ the Sun.

10. Earth's orbit is its _____path_____ around the Sun.

11. When Earth is _____close_____ to the Sun, the weather is warm.

12. When Earth is _____far_____ from the Sun, we have fall and winter.

Critical Thinking

13. Why does the weather change during the year?

The weather changes because Earth orbits the Sun
during the year. When Earth is close to the Sun, the
weather is warm. When it is farther away, the weather
is cooler.

_____

_____

_____

© Macmillan/McGraw-Hill

Why Seasons Happen

Circle the word that best tells about each picture.

1.

axis (orbit)

2.

(fall) winter

3.

spring (winter)

4.

fall (summer)

Name _____ Date _____

Why Seasons Happen

Fill in the blanks. Use the words from the box.

axis	orbit	seasons
fall	path	spring

You have learned that day and night are caused by Earth's rotation. Earth rotates on its

_____axis_____ . Do you ever wonder why

_____seasons_____ change?

As Earth rotates, it is also moving in a path around the Sun. This path is called Earth's

_____orbit_____ . It takes about 365 days for Earth to travel around the Sun once. The seasons

change when Earth's _____path_____ brings it closer or farther away from the Sun. When Earth is

closer to the Sun, we have warm _____spring_____ and summer weather. When Earth is far from the

Sun, we have cooler _____fall_____ and winter weather. What season would it be on the other side of the world right now?

Fun with the Seasons

✏ **Write About It**

Think about the seasons and the different activities you do throughout the year.

On a separate piece of paper, write a story about the activities you do in winter and in summer. Include details about how the seasons are different.

Getting Ideas

Fill in the chart with ideas about summer and winter.
Possible answer:

Winter	Summer	Both
weather gets colder; days get shorter; I go ice skating	weather gets warmer; days get longer; I go swimming	both seasons last three months; the weather changes; I hike in the woods

Planning and Organizing

Lisa wrote two sentences about winter and summer. Write Alike if the sentence shows how they are alike. Write Different if it shows how they are different.

1. _____Alike_____ Winter and summer are seasons.

2. _____Different_____ Winter can be very cold, and summer can be very hot.

Drafting

Write a sentence to begin your paragraph. Tell how you feel about winter and summer.

Possible answers: I like both summer and winter.

Now write your story on a separate piece of paper. Tell what you do in winter and summer. Tell how the seasons are different.

Revising and Proofreading

Lisa wrote some sentences. She made six mistakes. Find the errors. Then correct them.

I really like winter? I like to go ice-skating on the pond. I also like to go sleding. My favorite season is summer. It gets hot so I go to the beech every day with my friends. We look for shels. At night, we look at the stars and we try too find the Big Dipper. There are many activities to do in both seasons.

Now revise and proofread your writing. Ask yourself:

▶ Did I tell about what I like to do in winter?

▶ Did I tell about what I like to do in summer?

▶ Did I correct all mistakes?

© Macmillan/McGraw-Hill

The Moon and Stars

Use your book to help you fill in the blanks.

Why can we see the Moon from Earth?

1. The Moon does not shine like the _____Sun_____ .

2. We see the _____light_____ of the Moon because the Sun shines on it.

3. The Moon is many _____miles_____ away from Earth.

4. It is made of _____rock_____ and covered with dust.

5. The _____dust_____ helps the Moon look bright when the Sun shines on it.

Why does the Moon seem to change shape?

6. It takes the Moon about one _____month_____ to move around Earth.

7. The Moon's _____shape_____ seems to change every few days.

Name _____ Date _____

8. The different shapes we see during the month

are called _____ phases _____ of the Moon.

What are stars?

9. Stars are space objects made of hot _____ gases _____ .

10. Stars can have different _____ colors _____ and
sizes.

11. Some stars make _____ patterns _____ in the sky.

12. The Sun is a _____ star _____ that gives light
and heat to Earth.

Critical Thinking

13. Why can we see both the Moon and stars in the
night sky?

The stars give off their own light. The Moon shines

because light from the Sun shines on the Moon.

_____

_____

_____

The Moon and Stars

Complete each word.

1. The __m__ __o__ __o__ __n__ does not give off its own light.

2. The different shapes of the Moon are called __p__ __h__ __a__ __s__ __e__ __s__ .

3. It takes the Moon one month to go around __E__ __a__ __r__ __t__ __h__ once.

4. Stars look like tiny points of __l__ __i__ __g__ __h__ __t__ because they are so far away.

5. The Sun is the closest __s__ __t__ __a__ __r__ to Earth.

Name _____ Date _____

The Moon and Stars

Fill in the blanks. Use the words from the box.

gases	Moon's	phases	Sun
light	patterns	stars	

The Moon does not shine the way the Sun

does. We see the Moon because ____light____

from the Sun shines on the Moon. Even though it

looks different sometimes, the ____Moon's____

shape does not really change. The shapes of the

moon we see each month are called ____phases____ .

A star is an object in space made of hot

____gases____ . The ____Sun____ is the

closest star to the Earth. That is why it looks so

large. From Earth, other ____stars____ look like

tiny points of light. Some stars make

____patterns____ in the sky. Can you name any

star patterns?

The Solar System

Use your book to help you fill in the blanks.

What goes around the Sun?

1. Earth is a _____planet_____ .

2. Planets are huge _____objects_____ that move around the Sun.

3. Nine planets, their moons, and the _____Sun_____ make up our solar system.

4. Like _____Earth_____ , each planet in our solor system orbits the Sun.

5. The planets that are _____closer_____ to the Sun take less time to move around it.

What are the planets like?

6. The closest planet to the Sun is _____Mercury_____ .

Name _____ Date _____

7. Our planet has _____water_____ that we can drink and air that we can breathe.

8. Mars has a _____red_____ , rocky surface and two moons.

9. The smallest and coldest planet is called _____Pluto_____ .

Critical Thinking

10. Why do you think our group of planets is called a solar system?

Our group of planets is called a solar system because

they all move around the same Sun. The Sun is the

closest star to all of the planets.

_____

_____

_____

_____

_____

_____

The Solar System

Use the clues to solve the puzzle. Use the words from the box.

1. ___b___ Earth

2. ___g___ Mars

3. ___e___ Mercury

4. ___d___ planet

5. ___f___ Saturn

6. ___a___ solar system

7. ___c___ Venus

a. made of planets, Moons, and stars

b. where we live

c. the hottest planet

d. a huge object that moves around the Sun

e. the planet closest to the Sun

f. a planet with thick rings

g. the planet with a red, rocky surface

Name _____ Date _____

The Solar System

Fill in the blanks. Use the words from the box.

Jupiter	planet	solar system	Venus
Mars	rings	Sun	

There are nine planets in our solor system.

Earth is one of them. The _____Sun_____ is in

the center of the ____solar system____ . All of the

planets go around the _____Sun_____ . Each
planet is different.

Mercury and _____Venus_____ are closer to the

Sun than Earth. The planet _____Mars_____ has

a red, rocky surface. Neptune is a blue planet.

Saturn and Uranus both have _____rings_____

around them. _____Jupiter_____ is the largest
planet. Pluto is the farthest planet from the Sun.

Starry, Starry Night

Read the Reading in Science pages in your book. As you read, pay attention to important ideas. How did ancient sailors find the North Star? What did they do first? What did they do last? Write your ideas in the chart below.

First

Look for the Big Dipper.

↓

Next

Look for two stars on the edge of the Big Dipper.

Follow the stars up the outside edge of the Big Dipper to the Little Dipper.

↓

Last

Find the last star on the handle of the Little Dipper.

© Macmillan/McGraw-Hill

Name _____ Date _____

Write About It

Sequence. Long ago, sailors used star charts to find their way on the ocean. How do astronomers use star charts now?

Astronomers use star charts to guide telescopes on

Earth and in space.

I. What do you think is the main idea of this selection? Why?

The main idea is that stars have helped people find

their way for many years. I think so because the

details in the article give information about how

sailors and astronomers use constellations.

Earth and Space

Solve each riddle. Use the words in the box.

orbit	planets	rotation

1. _____planets_____

These great big rocks from 1 to 9, number 3 is home for us. The hottest one is called Venus!

2. _____orbit_____

Round and round, Earth and the Moon go, on a trip that makes the seasons switch!

3. _____rotation_____

It makes daytime here, and nighttime there, every 24 hours, every day, every year.

Name _____ Date _____

Label each picture. Use the words in the box.

solar system	Moon phases	Sun

1. _____Sun_____

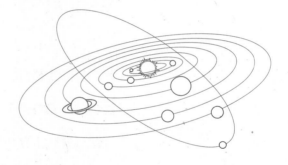

2. _____solar system_____

3. _____Moon phases_____

Popcorn Hop
by Stephanie Calmenson

Read the Unit Literature pages in your book.

 Write About It

Response to Literature

1. What makes the popcorn hop?

   Heat makes the popcorn hop.

   _____

2. How do you think popcorn got its name?

   Popcorn got its name because it makes a popping

   noise when it gets hot.

3. How do living things use heat?

   People use heat to cook and keep their homes warm.

   Plants and animals use heat from the Sun to stay

   warm.

Name _____ Date _____

Looking at Matter

Fill in the important ideas as you read the chapter. Write three facts about the properties of each kind of matter.

Matter is ___anything that takes up space and has mass___ .

What Are the Properties of Matter?		
Solid	**Liquid**	**Gas**
I. has its own shape	I. takes the shape of the container it is in	I. has no shape of its own
2. can float or sink	2. can use a measuring cup to measure liquids	2. includes oxygen, a gas that we breathe
3. can use a balance to measure solids	3. can flow	3. has volume and mass

© Macmillan/McGraw-Hill

Describing Matter

Use your book to help you fill in the blanks.

What is matter?

1. Matter is anything that takes up _____space_____ and has mass.

2. Some matter can be _____made_____ by people.

3. An object's mass is the amount of _____matter_____ it has.

4. Objects can be made of _____different_____ amounts of matter.

5. A _____balance_____ is used to measure and compare mass.

How can you describe matter?

6. Matter can be described by talking about its _____properties_____ .

7. A _____property_____ is how matter looks, feels, smells, tastes, or sounds.

8. Different _____kinds_____ of matter have different properties.

9. Matter can be _____living_____ or nonliving.

10. There are _____three_____ main kinds of matter: solids, liquids, and gases.

Critical Thinking

11. What are some ways that matter can be described? What do these ways tell you about matter?

Matter can be described as rough, smooth, wet,

round, or invisible. These different properties tell me

that matter can be anything that takes up space.

_____

_____

_____

_____

_____

_____

Describing Matter

What is the secret answer? Fill in the missing words and then fill in the answer by using the circled letters.

1. Matter can be t (h) i c k or thin.

2. Anything that takes up space and has mass is called m a (t) t e r .

3. Matter can be a s o l (i) d , liquid, or gas.

4. Matter can be natural or made by p (e) o p l e .

5. The amount of matter in an object is called m (a) s s .

6. A p r o p e r (t) y describes how matter looks, feels, smells, tastes, or sounds.

Q: What did the doctor say to the scientist?

A: W h a t i s t h e m a t t e r?

Describing Matter

Fill in the blanks. Use the words from the box.

balance	feel	gas	matter	smaller
describe	flexible	mass	property	

Matter is everywhere. Matter can be a solid, a

liquid, or a _____ gas _____ . Anything that takes up

space and has _____ mass _____ is matter. The

amount of _____ matter _____ in an object is called mass.

A _____ balance _____ can be used to measure and

compare the mass of objects. Sometimes, a _____ smaller _____

object has more mass than a larger object.

It is possible to _____ describe _____ matter by

talking about its properties. A _____ property _____ is a

way matter looks, feels, smells, tastes, or sounds.

Matter can be soft or it can be hard. Matter can be

_____ flexible _____ or stiff. It can also _____ feel _____

rough, smooth, or wet. Some matter is even invisible!

© Macmillan/McGraw-Hill

Solids

Use your book to help you fill in the blanks.

What is a solid?

1. A _____ solid _____ is one of three kinds of matter.

2. Solids have a _____ shape _____ of their own.

3. Like all matter, different solids have _____ different _____ properties.

4. Solids can be made from _____ materials _____ like wood, plastic, and metal.

5. They can feel smooth, rough, soft, or hard when you _____ touch _____ them.

How can we measure solids?

6. Many _____ tools _____ can be used to measure solids.

7. A _____ ruler _____ can be used to measure the width, length, or height of an object.

8. Rulers can be used to measure the lengths of objects in ___centimeters___ or inches.

9. A ___balance___ is used to tell how much mass something has.

10. To tell the difference between two objects, their measurements can be ___compared___ .

Critical Thinking

11. What will happen to a balance if you put a brick on one side and a feather on the other? Why?

The balance will tip toward the side with the brick.

This is because a brick has more mass than a feather.

_____

_____

_____

_____

_____

© Macmillan/McGraw-Hill

Solids

Circle the best answer.

1. Which solid is longer?

 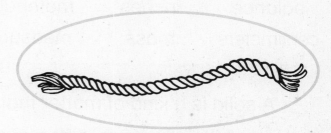

2. Which solid has less mass?

3. Which is softer?

 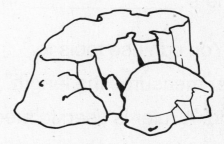

4. Which is smoother?

Solids

Fill in the blanks. Use the words from the box.

balance	inches	materials	properties	sink
centimeters	mass	measure	rough	

A solid is a kind of matter that has its own shape. Like all matter, different solids can be made of different _____materials_____ . Solids get their _____properties_____ from the materials they are made from. Solids can feel _____rough_____ , smooth, hard, or soft. Some solids float in water. Others _____sink_____ in water.

You can use tools to _____measure_____ solids. A ruler measures the length, width, and height of a solid. A ruler is used to measure lengths in units called _____centimeters_____ or in units called _____inches_____ . The amount of matter in a solid is called _____mass_____ . A _____balance_____ tells how much mass a solid has. Both methods of measurement can be used to form a more complete picture of objects.

Natural or Made by People?

Read the Reading in Science pages in your book. As you read, pay attention to important ideas. Summarize them in the chart below. Remember, when you summarize, you retell the most important ideas in the selection.

Summary

How are natural solids and manmade solids the same and different?

Natural and manmade solids can both be made

into objects that can be helpful to people. They are

different because people create manmade solids but

can only reshape a natural solid.

Idea #1	Idea #2	Idea #3
Natural solids come from earth. Manmade solids are made by people.	Natural solids can be cut or painted. Manmade solids are shaped using molds and colored using chemicals.	People can use natural solids and manmade solids in many ways.

Write About It

Summarize. How is a plastic chair made? Use the chart you made to write your answer.

First, people combine chemicals to make plastic.

They can use other chemicals to make different

colors. Then people use a mold in the shape of a chair

to shape the plastic.

_____

I. What are some plastic things in your classroom?

blocks, bookshelves, bulletin boards, dry-erase boards

Liquids and Gases

Use your book to help you fill in the blanks.

What is a liquid?

1. The opposite of _____liquid_____ matter is solid matter.

2. Unlike most solids, a liquid can take the shape of the _____container_____ it is in.

3. You can measure the _____volume_____ of a liquid by using a measuring cup.

4. Volume is a measure of the amount of _____space_____ something takes up.

What is a gas?

5. A _____gas_____ is like a liquid in many ways.

6. A gas has no _____shape_____ of its own.

7. A bubble is liquid with _____gas_____ inside it.

Name _____ Date _____

8. You can _____ the volume or the mass of a gas.

9. The _____ around us is made of many gases.

10. You can feel these gases moving on a _____ day.

11. We need a gas called _____ to breathe.

Critical Thinking

12. What solids, liquids, and gases do you use every day?

Possible answer: My toothbrush is a solid that I

use every day. I eat foods that are solids, such as

sandwiches and fruit. I drink liquids like water, milk,

and juice. I breathe oxygen, which is a gas.

_____

_____

_____

Liquids and Gases

Classify the words in the box based on their state of matter.

air	glass	ice	milk	pencil
apple	helium	juice	oxygen	water

Solids	Liquids	Gases
ice	water	oxygen
pencil	milk	helium
glass	juice	air
apple		

Liquids and Gases

Fill in the blanks. Use the words from the box.

air	containers	liquid	plants	three
breathe	gas	oxygen	solid	

We use matter every day. Our clothes, shoes,

breakfast, and even the _____ air _____ we

breathe are kinds of matter. There are _____ three _____

kinds of matter. A _____ solid _____ is a kind of

matter that has its own shape. A _____ liquid _____

is a kind of matter that does not have a shape of

its own. A _____ gas _____ is another kind of

matter that does not have its own shape.

Gases and liquids take the shapes of the _____ containers _____

they are in.

The air we _____ breathe _____ is made of many

gases. One of these gases in the air is called

_____ oxygen _____ . Animals and _____ plants _____

need oxygen to live. We cannot see gases but

they are all around us.

Fun with Water

Write About It

This girl is having fun in the water! Think of times that you have had fun in water. Draw and write about what you did.

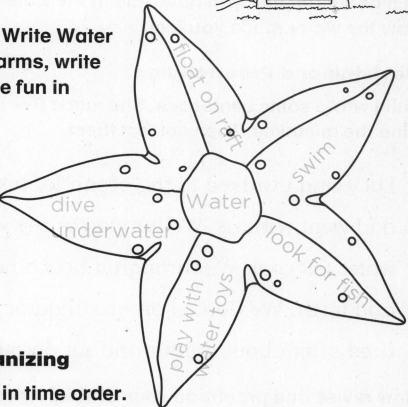

Getting Ideas

Look at the starfish. Write Water in the center. In the arms, write things you do to have fun in the water.

float on raft

swim

look for fish

Water

dive underwater

play with water toys

Planning and Organizing

Put these sentences in time order.

____3____ I jumped into the water.

____1____ I put on my bathing suit and packed some toys.

____2____ My mother and I walked to the beach.

Name _____ Date _____

Drafting

Write a sentence to begin your story. Use I to write about yourself.

Possible answer: Yesterday, I had a lot of fun at the

beach.

Now write your story on a separate piece of paper. Tell about fun that you have had in the water. Tell how the water made you feel.

Revising and Proofreading

Julia wrote some sentences. She made five mistakes. Find the mistakes. Then correct them.

Lucy and i walked to the ocean for a swim. His
                                              Her
dad went with us. We jumped in the weaves. The
                                        ew
water felt cool. We through a beach ball back
and forth. We floated on an alligator raft. We got
tired after about a hour and sat on our towels.
                        n

Now revise and proofread your writing. Ask yourself:

▶ Did I write about what I did in the water?

▶ Did I tell how I felt?

▶ Did I correct all mistakes?

© Macmillan/McGraw-Hill

Looking at Matter

Fill in the blanks. Use the words in the box.

balance	matter	solid
mass	property	volume

1. Anything that takes up space and has mass

   is _____matter_____ .

2. The amount of matter in an object is called

   _____mass_____ .

3. A _____balance_____ can be used to measure and
   compare mass.

4. The amount of space something takes up is

   called _____volume_____ .

5. A _____solid_____ has a shape of its own.

6. A _____property_____ is how matter looks, feels,
   smells, sounds, or tastes.

Write whether each fact describes a solid, a liquid, or a gas.

1. This kind of matter has a shape of its own.

   solid
   _____

2. It cannot be seen, but it is everywhere.

   gas
   _____

3. Water is an example of this kind of matter.

   liquid
   _____

4. Oxygen is an example of this kind of matter.

   gas
   _____

5. This can be made of plastic, metal, or wood.

   solid
   _____

6. This kind of matter can be measured by using a measuring cup.

   liquid
   _____

Changes in Matter

Using what you have learned from the chapter, fill in the blanks to tell how matter can change.

Physical Change	Chemical Change	Mixture
_____ cut	_____ changes from	_____ different combinations of
_____ bend	_____ heat	_____ solids, liquids, and gases
_____ fold	_____ changes from	_____ combining matter
_____ tear	_____ light	_____ solution

Change of State

_____ melt

_____ boil

_____ freeze

Name _____ Date _____

Matter Changes

Use your book to help you fill in the blanks.

What are physical changes?

1. Physical changes cause a _____change_____ in matter.

2. A physical change takes place when the size or shape of _____matter_____ changes.

3. The _____mass_____ of matter stays the same if its shape is changed.

4. When a piece of paper is folded or torn, a _____physical_____ change is taking place.

5. A change in _____temperature_____ can be a physical change, too.

6. When something gets _____wet_____ or dries, it may look and feel different, but it is only a physical change.

© Macmillan/McGraw-Hill

What are chemical changes?

7. During a _____chemical_____ change, one kind of matter changes into a different kind of matter.

8. When _____matter_____ goes through a chemical change, it may not be possible to change it back.

9. When wood is _____burned_____ in a fireplace, a chemical change is taking place.

10. Observing _____light_____ and feeling _____heat_____ and cold are clues that a chemical change may be occuring.

Critical Thinking

11. Think about a piece of bread. How can you make a physical change to the bread? How can you make a chemical change?

Possible answer: I can cut the bread to make a

physical change. I can also tear, smash, and flatten

the bread to make a physical change. I can toast the

bread to make a chemical change.

Matter Changes

Identify each description as a physical change or a chemical change.

1. An iron screw rusts in the rain.

chemical change

2. A piece of paper is folded.

physical change

3. A rock breaks down into soil.

physical change

4. Water freezes and turns into ice.

physical change

5. A peach turns brown.

chemical change

6. A ball gets wet.

physical change

7. A slice of cheese melts.

physical change

8. An egg is fried.

chemical change

Matter Changes

Fill in the blanks. Use the words from the box.

burns	mass	rusts
chemical change	matter	temperature
fold	physical change	

Matter changes every day. A _physical change_ takes place when the size or shape of matter changes but not the type of matter. When you _fold_ paper, you are making a physical change. When only the shape of an object changes, its _mass_ stays the same. When the _temperature_ of water changes, it can freeze or boil. These are physical changes, too.

You can also make a _chemical change_ to matter. A chemical change happens when _matter_ changes into a different kind of matter. When matter _burns_, it can not change back to its original form. When iron _rusts_, it changes color and feels different. These are chemical changes at work.

© Macmillan/McGraw-Hill

Changes of State

Use your book to help you fill in the blanks.

How can heating change matter?

1. Heat can change _____ matter _____ in different ways.

2. When a solid gets enough _____ heat _____, it melts.

3. When something melts, it changes from a _____ solid _____ to a liquid.

4. When heat is added to ice, it turns into _____ liquid _____ water.

5. Different solids can _____ melt _____ at different temperatures.

6. Some liquids _____ boil _____ when they get enough heat.

7. When liquid water boils, it _____ evaporates _____, or changes into a gas.

8. This gas is called _____ water vapor _____ .

How can cooling change matter?

9. When you _____cool_____ matter, you take heat away from it.

10. A gas can _____condense_____ when it is cooled.

11. When a _____gas_____ condenses, it changes into a liquid.

12. When _____liquids_____ lose enough heat, they freeze.

13. When matter _____freezes_____, it changes from a liquid to a solid.

Critical Thinking

14. Explain how you can make an ice cube change from a solid to gas.

Possible answer: First, I can leave the ice cube out to

melt. It will turn from a solid to a liquid. Then, I can

heat the liquid until it boils. The water will evaporate

and turn into a gas.

_____

_____

Name _____ Date _____

Changes of State

Solve the riddles and fill in the puzzle.

Down

1. I keep my shape when I'm cool. If it gets too warm, I melt.   ice

2. You can add me or take me away to change matter.   heat

4. This happens when I get very cold.   freeze

6. When I start out very hot and then become cool, I turn into liquid.   gas

Across

3. This is what I do when 6 Down happens.   condense

5. This is how I turn solids into liquids.   melting

7. This is how I go into the air when I'm boiling.   evaporate

© Macmillan/McGraw-Hill

Changes of State

Fill in the blanks. Use the words from the box.

condense	heat	solid
evaporate	liquid	temperatures
freeze	melt	water vapor

There are three main states, or forms, of matter.

The three main states are _____solid_____ , liquid,

and gas. Some solids _____melt_____ when they

get enough heat. When something melts, it

changes from a solid to a _____liquid_____ . That is

what happens when an ice cube melts. Different

solids must be heated to different _____temperatures_____

in order to melt. When water boils, it will _____evaporate_____ ,

or turn into a gas. This gas is called _____water vapor_____ .

When _____heat_____ is taken away from matter,

it can change. Gases _____condense_____ when they

are cooled. When you _____freeze_____ water, it

turns into a solid. Different liquids freeze at

different temperatures.

Colorful Creations

Read the Reading in Science pages in your book. Write inferences based on the statements in the "What I Know" column. Write your inferences on the chart.

What I Know	What I Infer
Most crayons are made of wax. Colored wax is melted into a liquid.	The liquid will be poured into a mold that is shaped like a crayon.
The crayon mold is cooled with cold water.	The hot liquid wax will become solid when it is cool.
A machine packs the crayons into boxes.	People will buy the crayons when they are sent to stores.

© Macmillan/McGraw-Hill

✏ Write About It

Predict. What do you think would happen if the mixture of wax was poured into a mold shaped like a square? Explain your answer.

If the mold were shaped like a square, then the melted wax would cool and harden into that shape. I know this because liquid takes the shape of any container it is poured into, and once it becomes a solid, it keeps that shape.

What two states of matter are used to make crayons?

solids and liquids

How do you think different colored crayons are made?

Different dyes are added to the wax to make different-colored crayons.

© Macmillan/McGraw-Hill

Name _____ Date _____

Mixtures

Use your book to help you fill in the blanks.

What are mixtures?

1. When two or more things are put together, the

   result is called a _____mixture_____ .

2. Mixtures can have different ____combinations____ of solids, liquids, and gases.

3. Some mixtures can be picked _____apart_____ .

Which mixtures stay mixed?

4. A mixture that is difficult to take apart is called a

   _____solution_____ .

5. When salt is added to water, the salt

   _____dissolves_____ and mixes with the water.

6. Sand and water _____do not_____ make a solution.

© Macmillan/McGraw-Hill

How can you take mixtures apart?

7. Some mixtures are _____easy_____ to take apart. Other mixtures are more difficult.

8. A _____filter_____ can be used to separate sand from water.

9. A _____magnet_____ can be used to separate iron from sand.

10. To take out salt from salt water, a process called _____evaporation_____ is used.

Critical Thinking

11. Suppose you had a mixture of water and pebbles. How could you take apart the mixture?

   Possible answer: I could use a filter to take out the

   pebbles from the water, I could pick out the pebbles

   with my hands, or I could leave the mixture out and

   let the water evaporate.

   _____

   _____

   _____

   _____

Name _____ Date _____

Mixtures

Write whether you would need to use a magnet, a filter, evaporation, or your hands in order to take apart each mixture listed below. Some mixtures can be taken apart in more than one way.

1. salt water

_____evaporation_____

2. water and sand

_____filter, evaporation_____

3. iron nails and sand

_____hands, magnet_____

4. raisins and cornflakes

_____hands_____

5. iron screws and plastic beads

_____hands, magnet_____

6. pennies and nickels

_____hands_____

7. blue paper and white paper

_____hands_____

8. water and seashells

_____filter, hands,_____
_____evaporation_____

© Macmillan/McGraw-Hill

Mixtures

Fill in the blanks. Use the words from the box.

dissolves	filter	magnet	separate
evaporation	liquids	mixture	solution

Have you ever made a collage? When you glue

pieces of paper together, you make a _____mixture_____ .

A mixture can be any combination of solids, _____liquids_____ ,

and/or gases. Some mixtures can be _____separated_____

by their parts.

When salt and water are mixed together, a

_____solution_____ is made. The salt cannot be seen

because it _____dissolves_____ in the water. The mixture

can be taken apart by using _____evaporation_____ . The

water will evaporate and the salt will be left behind.

To separate water and sand, a _____filter_____

can be used. To separate iron and sand, a _____magnet_____

can be used. You can separate some mixtures by

using your hands.

Name _____ Date _____

Writing a Recipe

Write About It

You can write a recipe. Explain how you would use some of this fruit to make a fruit salad. Explain why it is a mixture.

Getting Ideas

Look at the illustration. What kinds of fruit do you see? Think about how you would make a fruit salad.

What kinds of fruit would you want to put in a fruit salad? List them below.

apples, bananas, grapes, blueberries, strawberries
_____

Planning and Organizing

Put the steps in the correct order.

_____4_____ Mix the fruit together.

_____2_____ Wash the fruit and put it on the cutting board.

_____1_____ Get a bowl and a cutting board.

_____3_____ Cut up each fruit. Put the fruit in the bowl.

Drafting

Write a sentence to begin your recipe. Tell what the recipe is for.

Possible answer: This recipe shows how to easily make a

fruit salad.

Now write the recipe on a separate piece of paper. Put the steps in order. At the end, tell why it is a mixture.

Revising and Proofreading

Use the words in the box to fill in the blanks.

Finally	First	Next	Second	Then

_____First_____ , I put a big bowl on the counter.

_____Second_____ , I got a spoon. _____Next_____ ,

I put cut-up apples and bananas in the bowl. _____Then_____ ,

I added grapes, blueberries, and strawberries.

_____Finally_____ , I mixed everything together.

Now revise and proofread your writing. Ask yourself:

▶ Did I write the steps in order?

▶ Did I explain why it is a mixture?

▶ Did I correct all mistakes?

Name _____ Date _____

Changes in Matter

Fill in the blanks. Use the words in the box.

chemical change	evaporation	melts
condenses	freezes	solution

1. When matter _____melts_____ , it changes from a solid to a liquid.

2. A process called _____evaporation_____ can be used to separate salt from water.

3. A _____solution_____ is a mixture that is difficult to separate.

4. When matter _____condenses_____ , it changes from a gas to a liquid.

5. When water _____freezes_____ , it changes from a liquid to a solid.

6. When a slice of bread is toasted, a _____chemical change_____ occurs.

© Macmillan/McGraw-Hill

Draw a line from each picture to the sentence that describes it.

1.

a. Salt dissolves in water to make a solution.

2.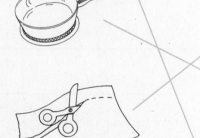

b. When a physical change takes place, matter changes shape.

3.

c. Evaporation is when matter changes from a liquid to a gas.

4.

d. After a chemical change takes place, matter may look and smell different than before.

5.

e. When matter melts, it changes from a solid to a liquid.

Chapter 10 • Changes in Matter
Reading and Writing

Echolocation

Read the Unit Literature pages in your book.

Write About It

Response to Literature

1. Why do you think that bats use echolocation?
Use the article to tell how you know.

I think they use echolocation because they can hear

better than they can see. I know this because the

article tells me that bats live in dark caves and cannot

see very well.

2. What other animals do you think use
echolocation?

Possible answer: I think that dolphins, who live in the

dark oceans, use echolocation.

3. Have you ever used sound to find something?
Write about it.

Possible answer: I follow the sound of my mother's

voice when she calls me for dinner.

© Macmillan/McGraw-Hill

How Things Move

Fill in the important ideas as you read the chapter.

How do things move?		
Where things move	**What makes things move**	**Ways things move**
position	forces	simple machines
distance	friction	magnets
speed	gravity	

Name _____ Date _____

Position and Motion

Use your book to help you fill in the blanks.

What are position and motion?

1. You can use _____position_____ words to describe an object's location.

2. Position is the _____place_____ where something is.

3. Above, _____below_____, left, and right are all position words.

4. When an object _____moves_____, it changes position.

5. When an object is moving, it is in _____motion_____ .

6. You can _____compare_____ the position and motion of objects.

What is speed?

7. Speed is a ____measure____ of how quickly an object changes its position.

8. Some objects and ____living____ things move quickly.

9. Speed shows the ____time____ it takes to move a certain distance.

10. Distance is a measure of how ____far____ something moves.

Critical Thinking

11. Scientists use tools to measure objects. What kind of tool could you use to measure the distance an object has moved? How?

Possible answer: I think I could use a ruler or a meter stick to measure distance. I think so because these tools measure length. To measure distance, I can use the ruler to see how much space there is between the object's first position and its next position.

Name _____ Date _____

Position and Motion

Describe the position of the bowl below in as many ways as you can.

Possible answer: The bowl is on the table. It is above the

dog. It is below the painting. It is to the right of the glass.

It is to the left of the milk.

Position and Motion

Fill in the blanks. Use the words from the box.

compare	motion	space
left	object	stopwatch
measure	position	

How do you know where something is? We use

words like above, below, ____left____ , and

right to describe where things are. When you

describe an object's ____position____ , you tell

where it is. To tell the position of an object, you

can ____compare____ it to another object.

Objects do not always stay in the same place.

When an ____object____ moves, its position

changes. This is called ____motion____ . Speed

is a ____measure____ of how quickly an object

moves from one position to another. The ____space____

between the two positions is called distance. You

can use a ____stopwatch____ to measure speed.

You can use a tape measure to measure distance.

© Macmillan/McGraw-Hill

Name _____ Date _____

Forces

Use your book to help you fill in the blanks.

What makes things move?

1. It takes a _____ or a pull to make
   something move.

2. A push or pull is a _____ .

3. To push something, you move it _____
   you.

4. To pull something, you move it _____
   you.

What are some forces?

5. When you throw a ball in the air, _____
   pulls it back to Earth.

6. Gravity is a force that _____ things
   to Earth.

7. One _____ of gravity is weight.

8. _____ is how much force it takes to
   pull something to Earth.

How can forces change motion?

9. Forces can make things _____speed_____ up,
slow down, or change direction.

10. Sometimes, objects _____rub_____ together
when they move.

11. When this happens, a force called
_____friction_____ slows down the objects.

Critical Thinking

12. Do you think gravity is important? Why or
why not?

Possible answer: Yes. I think gravity is important

because it helps keep things on the surface of Earth.

Without gravity, we might float off into outer space.

_____

_____

_____

_____

_____

Forces

Answer each riddle. Then find each word in the word search.

1. I am a force that slows down moving things.

What am I? _____friction_____

2. I am a force that pulls things to Earth.

What am I? _____gravity_____

3. To put an object in motion, you must use me.

What am I? _____force_____

4. I am the amount of force Earth pulls on an

object. What am I? _____weight_____

5. To move an object closer to you, you must use

me. What am I? _____pull_____

f	r	i	c	t	i	o	n	d	w
o	l	m	s	h	i	e	h	g	e
r	g	r	a	v	i	t	y	c	i
c	a	t	v	m	p	s	t	u	g
e	m	n	x	y	r	l	m	e	h
p	u	l	l	n	z	c	b	o	t

© Macmillan/McGraw-Hill

Forces

Fill in the blanks. Use the words from the box.

amount	down	pull
away	force	push
direction	gravity	

How do you move things? Think about the last time you threw a ball. You used a _____force_____ to move the ball. A force is a _____push_____ or pull that makes objects move. When you _____pull_____ an object, you move it closer to you. When you push an object, it moves _____away_____ from you.

You can use forces to speed up or slow _____down_____ an object. Friction is a force that slows some things down. Forces can even change the _____direction_____ of an object's motion. The force that pulls objects to Earth is called _____gravity_____ . The _____amount_____ of force that gravity pulls down on an object is called weight. People use forces every day.

Meet Hector Arce

Read the Reading in Science pages in your book.
As you read, keep track of what happens and why.
Record the causes and effects you read about in the
chart below. Remember, a cause is why something
happens. An effect is the thing that happens.
Sometimes, one cause can have many effects.

Cause	Effect
Gravity	It keeps living things and objects on Earth.
Gravity	It pulls together huge clouds of gas and dust to form stars.
Gravity	It makes the centers of stars hot enough to glow.

© Macmillan/McGraw-Hill

Use the words in the box to retell what you learned about the effects of gravity.

dust	gas	hot
force	gravity	stars

The _____force_____ that pulls objects toward

Earth is called _____gravity_____ . It keeps all living

things and objects on Earth as the planet spins.

Gravity also pulls on other planets and on moons.

It can even cause _____stars_____ to form.

Gravity pulls together clouds of _____gas_____

and _____dust_____ to make stars. Inside these

stars, gravity makes them so _____hot_____

that they glow in the night sky.

Write About It

Cause and Effect. What causes stars to form?

The force of gravity pulls huge clouds of gas and dust

together and then causes them to grow hot enough

to glow in the sky.

© Macmillan/McGraw-Hill

Using Simple Machines

Use your book to help you fill in the blanks.

What are levers and ramps?

1. A __simple machine__ is a tool that can change the strength of a force.

2. A __lever__ is a simple machine with a bar that moves on a stationary fulcrum.

3. This machine can change how much force is needed for a __push__ so you can move heavy things.

4. A seesaw and a __crowbar__ are kinds of levers.

5. A __ramp__ is another kind of simple machine that can help you move things.

6. A ramp has a __straight__, slanted surface.

What are other simple machines?

7. People use simple machines like axles and

_____pulleys_____ every day.

8. An _____axle_____ is a bar that is connected to the center of a wheel.

9. A simple machine made of a rope that moves

around a _____wheel_____ is called a pulley.

10. Pulleys make it easier to _____pull_____ things.

Critical Thinking

11. Where have you seen ramps? Why are these and other simple machines useful?

Possible answer: I have seen ramps in front of stores

and buildings. Ramps and other simple machines

can be used to move heavy things from one place to

another with less force.

_____

_____

Name _____ Date _____

Using Simple Machines

Identify the simple machine in each picture.

1.

   ramp

2.

   wheel and axle

3.

   lever

4.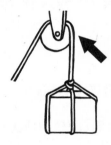

   pulley

5.

   wheel and axle

Using Simple Machines

Fill in the blanks. Use the words from the box.

axle	lever	simple machine
force	pulley	
fulcrum	ramp	

Tools help people change the _____ force _____
used on an object. What tools do you use every
day? Sometimes, objects are too heavy to lift or
move on our own. A ___ simple machine ___ is a tool
that can change the size of a force. A _____ lever _____
is a bar that moves on a point that stays still. This
point is called a _____ fulcrum _____ . People use
this tool to lift heavy things. A _____ ramp _____
is used to move things from one place to another.
We can push objects on its slanted surface.

Cars and bikes have wheels that help them
move. An _____ axle _____ is a bar connected to
the center of the wheel. A _____ pulley _____ has a
rope that moves around a wheel. This tool helps
people change the direction of an object.

Name _____ Date _____

Slip and Slide

 Write About It

Explain why penguins can slide on the ice.
Think about what you learned about forces.
Make sure to explain why ice is slippery.

Getting Ideas

Brainstorm a list of facts about penguins, and write them in the chart below.

Penguin Fact Sheet
flippers
layer of fat
short legs
black and white feathers

Planning and Organizing

Zina wrote four sentences. Write Yes if the sentence is a penguin fact. Write No if it is not.

1. ____Yes____ Penguins are birds that have webbed feet.

2. ____Yes____ Penguins have black and white feathers.

3. ____No____ Penguins can fly.

4. ____Yes____ Penguins have short legs.

Name _____ Date _____

Writing in Science

Drafting

Write your own topic sentence to begin your paragraph. Tell your main idea about penguins.

Possible answer: The penguin's special body helps it to slide on the ice.

Now write about penguins on a separate piece of paper. Start with your main idea. Explain how they slide on the ice. Tell which body parts help them move.

Revising and Proofreading

Zina wrote a paragraph. She made five mistakes. Find the mistakes. Then correct them.

Penguins slide on their bellys. They use their feet and flipers. Their feet push them forward. There flippers balance them. When they glide, the ice under them melts. This makes the ice slipperie. They can glide a few miles an our. Gliding takes less energy than walking.

Now revise and proofread your writing. Ask yourself:

▶ Did I follow all instructions?

▶ Did I correct all mistakes?

© Macmillan/McGraw-Hill

Name _____ Date _____

Exploring Magnets

Use your book to help you fill in the blanks.

What do magnets do?

1. Magnets use _____force_____ to attract some objects.

2. Magnets can pull objects without _____touching_____ them.

3. A _____magnet_____ can attract objects made of iron, nickel, or steel.

4. Strong magnets can _____pull_____ objects that are far away.

5. Magnets can pull objects that contain _____nickel_____ or steel.

6. Magnets cannot pull objects made of _____wood_____ or plastic.

What are poles?

7. The _____poles_____ are the two ends of a magnet.

8. All magnets have a north pole and a ____south_____ pole.

9. The ____north_____ pole and south pole are opposites.

10. The north pole of one magnet and the south

    pole of another magnet will ____attract_____ each other.

11. Two like magnetic poles will ____repel_____ one another.

Critical Thinking

12. How do people use magnets?

    Possible answer: People can use magnets to pick

    up objects made out of iron. They can use magnets

    to stick objects together. People can lift large iron

    objects with large magnets.

    _____

Name _____ Date _____

Exploring Magnets

**If a magnet will attract the object, write Will attract.
If a magnet will not attract the object, write Will not attract.**

1.

paper clip

Will attract

4.

pencil

Will not attract

2.

screw

Will attract

5.

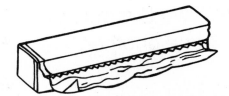

aluminum foil

Will not attract

3.

penny

Will not attract

6.

nail

Will attract

Exploring Magnets

Fill in the blanks. Use the words from the box.

attract	magnet	north
iron	nickel	south

It is possible to move objects without even touching them. A ___magnet___ can make some things move. It uses force to ___attract___ , or pull, some objects. It can pull objects that contain ___iron___ , like paper clips and screws. It can also pull objects made out of ___nickel___ or steel. A magnet can not attract things made out of wood or plastic.

Every magnet has two poles. If the ___north___ pole of one magnet is put next to the south pole of another magnet, the two magnets will attract. If the ___south___ pole of one magnet is put next to the south pole of another, the two magnets will repel. Magnets are powerful!

Name _____ Date _____

How Things Move

Fill in the blanks. Use the words in the box.

friction	lever	position
gravity	poles	simple machine

1. A _simple machine_ is a tool that can change the size of a force.

2. A force that slows down moving things is called _friction_ .

3. Every magnet has two _poles_ .

4. A _lever_ is a simple machine that helps people lift heavy things.

5. You can tell the _position_ of an object by comparing it to another object.

6. The force that pulls things toward the ground is called _gravity_ .

Complete the sentences. Then fill in the puzzle.

Down

1. When you ___push___ something, you move it away from you.

3. A ___ramp___ is a simple machine with a straight surface that is slanted.

5. The amount of force that pulls an object down toward Earth is called its ___weight___ .

Across

2. An ___axle___ is a bar that is connected to the center of a wheel.

4. The point on a lever that stays still is called the ___fulcrum___ .

Name _____ Date _____

Using Energy

**Fill in the important ideas as you read the chapter.
Use the words in the box to fill in the first row. Use
your own ideas to fill in the second row.**

heat	light	sound

light	sound	heat

Using Energy

How We Use It	How We Use It	How We Use It
to see things	to hear things	to keep warm
to see in color	to talk to each	to cook food
	other	

© Macmillan/McGraw-Hill

Heat

Use your book to help you fill in the blanks.

What is heat?

1. Energy makes _____ matter move or change.

2. Heat is energy that can change the _____ state of matter.

3. Heat can _____ melt solids and turn liquids into gases.

4. The _____ Sun gives Earth most of its heat.

5. We can also get heat _____ energy from other things.

6. Something that gives off heat energy when it is burned is _____ fuel .

7. Heat energy can also come from _____ motion .

What is temperature?

8. We can tell how hot or cold something is by

measuring its ___temperature___ .

9. Thermometers have a special ___liquid___
inside of them.

10. When the temperature is ___warm___ , the
liquid goes up.

11. When the temperature is cool, the liquid goes

___down___ .

Critical Thinking

12. What are three sources of heat energy?
How do we measure this energy?

___Heat energy comes from the Sun. It can also come___

___from fuels, like gas or wood. Heat energy also___

___comes from motion. We measure this energy with a___

___thermometer that tells us the temperature.___

_____

_____

Heat

Read each sentence. Write TRUE if the sentence is true. Write NOT TRUE if the sentence is false.

1. _____TRUE_____   Heat energy can change the states of matter of some objects.

2. _____NOT TRUE_____   Heat can turn a gas into a solid.

3. _____NOT TRUE_____   Most heat energy comes from the Moon.

4. _____TRUE_____   Gas, oil, wood, and coal are all types of fuel.

5. _____TRUE_____   Temperature is a measure of how hot or cold something is.

6. _____NOT TRUE_____   Thermometers measure how fast someone is running.

Name _____ Date _____

Heat

Fill in the blanks. Use the words from the box.

coal	heat energy	motion
fuel	matter	temperature

There are many elements of energy. Energy

makes _____matter_____ move or change. The

Sun gives _____heat energy_____ to Earth. Heat energy

keeps us warm.

Not all heat energy comes from the Sun. Gas, oil,

wood, and _____coal_____ give off heat energy.

Things that give off heat when burned are called

_____fuel_____ . You can make heat energy, too!

When you rub your hands together quickly, the

_____motion_____ makes heat energy.

A measure of hot and cold is called _____temperature_____ .

A thermometer is a tool that people use to measure

temperature.

Sound

Use your book to help you fill in the blanks.

What makes sound?

1. Another kind of energy we use every day is
_____sound_____ .

2. When objects _____vibrate_____ , they give off
sound energy.

3. *Vibrate* means "to move _____back_____ and
forth quickly."

4. When your _____eardrum_____ vibrates, you hear
sound.

5. Your _____brain_____ helps you figure out what
you are hearing.

How are sounds different?

6. Some sounds are _____soft_____ and some
sounds are loud.

7. Soft sounds have less energy than
_____loud_____ sounds.

8. Some sounds have a higher _____pitch_____ than other sounds.

9. Pitch is how high or _____low_____ a sound is.

What do sounds move through?

10. Sound can _____move_____ through air.

11. Sound energy can even move through _____solids_____ and many liquids!

Critical Thinking

12. How do we hear sound? How are sounds different?

Sounds can travel through air, liquids, or solids before

they go to our eardrums. Sounds can be loud or soft,

and can be high or low in pitch.

_____

_____

_____

_____

_____

Sound

Describe what each picture shows about sound.

1.

The alarm clock is vibrating.

2.

The eardrum is the part of the body we use to hear sounds.

3.

Pitch is how high or low a sound is. If you snap a long, loose string, it makes a low pitch.

Name _____ Date _____

Sound

Fill in the blanks. Use the words from the box.

eardrum	liquids	sound
energy	pitch	vibrate

Did you know that we can hear a kind of

energy? The kind of energy that we can hear is

_____sound_____ . Sound energy is made when

objects _____vibrate_____ . Sound can travel

through air. Sound can also travel through solids

and _____liquids_____ . The closer you are to a

sound, the louder it will be.

How do we hear these sounds? The part of our

body we use to hear sounds is the _____eardrum_____ .

It sends messages to our brain about what sound

we heard. Not all sounds are the same. A whisper has

less _____energy_____ than a shout. The _____pitch_____

is how high or low a sound is. Imagine a guitar's strings.

The tighter the strings are, the higher the pitch is. There

are many different sounds.

Sound Off!

 Write About It

Describe the pitch and volume of a sound you hear every day. How do we use sounds? Why are sounds important?

Getting Ideas

Choose a sound you hear every day. Write it in the center ovals. In the outer ovals, write words that describe that sound.

Possible answer:

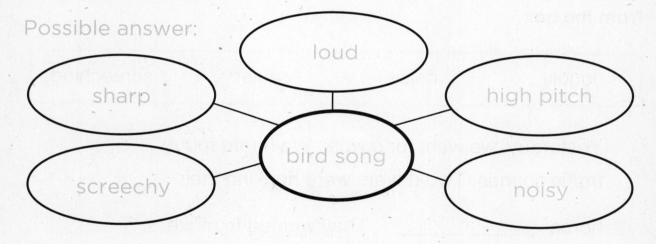

loud

sharp

high pitch

bird song

screechy

noisy

Planning and Organizing

Circle the descriptive words in these sentences.

1. The brown sparrow sang loudly.

2. The little sparrow sang a pretty song.

© Macmillan/McGraw-Hill

Name _____ Date _____

Drafting

Write a sentence to begin your paragraph that tells an important idea about a sound you hear every day.

Possible answer: Every day, I hear the noise of traffic.

Write about the sound on a separate piece of paper. Remember to use descriptive words.

Revising and Proofreading

Pedro wrote a paragraph. He did not use any describing words. Fill in the blank spaces with words from the box.

loudly	noisy	quiet	screeching

Yesterday, we went for a walk. We heard many traffic sounds. Two drivers were honking their

horns _____loudly_____ . They wanted to make sure a boy on a bike saw them. A car stopped at

a red light. It made a _____screeching_____ sound. Then two fire engines went zooming past us.

The traffic sounds were so _____noisy_____ .

There was not one _____quiet_____ place in the city.

© Macmillan/McGraw-Hill

Light

Use your book to help you fill in the blanks.

What is light?

1. Did you know that _____light_____ energy helps you see things?

2. Some light comes from _____lightbulbs_____ and flashlights.

3. Most light on Earth comes from the _____Sun_____ .

4. Light _____reflects_____ off of objects and goes into our eyes to help us see.

5. The dark area made when something is blocking light is called a _____shadow_____ .

6. Some _____solid_____ objects can block light and make shadows.

How do we see color?

7. White light is really a mix of different _____colors_____ of light.

8. When light _____bends_____ , we can see the colors of the rainbow.

9. A _____prism_____ is a tool that helps to bend light.

10. A _____filter_____ is a tool that blocks some colors of light.

Critical Thinking

11. Why is light important? How many kinds of energy does the Sun give to Earth?

Light is important because we need light to see

things. The Sun gives heat and energy to Earth.

_____

_____

_____

_____

_____

_____

_____

Light

Fill in the blanks. Use the words from the box.

colors	eyes	prism	reflects
energy	light	rainbow	

1. Light is a mix of _____ colors _____ .

2. My _____ eyes _____ are important tools that let me see the world around me.

3. Heat, sound, and light are all kinds of _____ energy _____ .

4. To see things, we must have _____ light _____ .

5. A _____ prism _____ can bend light.

6. When light _____ reflects _____ off objects and enters our eyes, we can see those objects.

7. If you shine light through a prism, you can see a _____ rainbow _____ .

© Macmillan/McGraw-Hill

Light

Fill in the blanks. Use the words from the box.

colors	prism	shadow
light	reflects	solid

You would not be able to see anything if there were no light. Some sources of ____light____ are the Sun, lightbulbs, and flashlights. We see objects because the light from these sources ____reflects____ off of objects around us. A ____shadow____ is a dark area that light does not reach. Light cannot pass through some ____solid____ objects. Light can pass through clear objects such as, glass.

Light is a mix of all ____colors____ . An object that makes light bend is called a ____prism____ . When light bends, it separates into the different colors of the rainbow.

© Macmillan/McGraw-Hill

Exploring Electricity

Use your book to help you fill in the blanks.

What is current electricity?

1. Electricity is a kind of _____energy_____ that gives off light and heat.

2. Electricity that moves in a path is called ___current electricity___ .

3. We call this path a _____circuit_____ .

4. Current electricity can come from _____batteries_____ or from outlets.

5. Power _____plants_____ make electricity that connects to wall outlets in homes.

What is static electricity?

6. The kind of energy that helps things stick together is called ___static electricity___ .

7. Pieces of matter push toward or pull from each

   other when they have a _____charge_____ .

8. A charge can build up on one object and then

   _____jump_____ to another object.

9. This is how _____lightning_____ works.

10. Charges build in storm _____clouds_____ and
    then jump to the ground.

Critical Thinking

11. How are a flashlight and lightning similar?
    How are they different?

    Both are lit by electricity. The flashlight is lit by

    current electricity. Lightning is made from static

    electricity.

    _____

    _____

    _____

    _____

Exploring Electricity

Match each picture to the word that tells about it.

1. ____ c ____ current electricity   **a.**

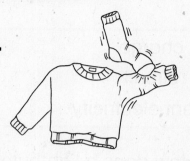

2. ____ d ____ circuit   **b.**

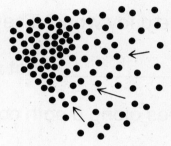

3. ____ a ____ static electricity   **c.**

4. ____ b ____ charge   **d.**

Name _____ Date _____

Exploring Electricity

Fill in the blanks. Use the words from the box.

charge	energy	static electricity
circuit	flow	
current electricity	outlets	

How does a lightbulb light up? How do batteries make a toy work? They need a kind of energy called current electricity _____ that moves in a path. The electricity moves along a path called a ____circuit_____ . In order for the electricity to ____flow_____ , the circuit needs to be closed. Current electricity can be changed into heat, light, or sound ____energy_____ .

It can come from batteries, ____outlets_____ in the wall, or other sources.

There are other kinds of electricity, too. A kind of energy made by tiny pieces of matter is static electricity _____. When tiny pieces of matter attract or repel each other, they have a ____charge_____ . Lightning is an example of static electricity. Electricity is everywhere!

It's Electric

Read the Reading in Science pages in your book. As you read, keep track of what happens and why. Record the causes and effects you read about in the chart below. Remember, a cause is why something happens. An effect is the thing that happens.

Cause	Effect
coal, oil, wind, water, or nuclear reactions	energy
The turbine turns a magnet inside the generator.	The generator creates electricity.
You flip a switch in your home.	Electricity flows from the power plant through power lines to your home.

Name _____ Date _____

Energy is needed to make electricity. Where can that energy come from?

The energy to make electricity can come from burning

coal or oil, flowing water or wind, or even nuclear power.

Where does energy come from in your community? Ask an adult to help you find out!

Possible answer: In my community, energy comes from

natural gas.

 Write About It

Cause and Effect. How does electricity help make your life easier?

Electricity makes it easier and faster to turn on lights,

travel from place to place, find information, and even

to cook food.

Using Energy

**Match the vocabulary word on the left with the letter
of the phrase that describes it.**

1. ____d____ current
electricity

a. to move
backward and
forward quickly

2. ____a____ vibrate

b. a path for electricity

3. ____b____ circuit

c. energy that can change
the state of matter

4. ____e____ charge

d. energy that can be
changed to heat, light,
or sound energy

5. ____c____ heat

e. a force that makes
tiny pieces of matter
sometimes attract or
repel each other

Name _____ Date _____

Circle the word that best tells about each picture.

1.

charge   (circuit)   vibrate

2.

(reflect)   prism   shadow

3.

energy   friction   (static electricity)

using after the DMV took away her license for good. It makes her mad she can't recall the year it happened; she likes to think she's still sharp, although with snow and sleet sliding down the windshield, she has to admit her vision is shot. She'll write that down later and maybe even read it out loud when her senior journaling class gets back together again in two weeks. A certain amount of honesty is important. *The journal's about you*, the twenty-five-year old instructor, Felicia, who doesn't like that Ella writes in the third person, had said. *Couldn't you switch over?*

No she could not. The girl's clicking tongue studs were starting to annoy her.

Dusky violet clouds in a dead winter sky turn the bronze archways of Michigan Central Station slime green. Parking in front of the monstrosity, she lets the motor run and surveys broken glass in the old railroad station's entryway doors. There are no other cars on the street but she's not afraid, as most of her neighborhood in southwest Detroit, where she's lived on Gilbert Street for over forty-five years, is empty too. Of course if someone saw her parked in front of Michigan Central Station, the very same train depot to which she and Arthur returned from California after the war, they might think her memory is gone. Thank you very much, it is not.

Since she's the only able-minded woman left in Detroit who built the big ships, she's been asked to give a little speech about what it was like in those days.

When Akin told her he'd found the barbed wire surrounding the building cut, she knew immediately he'd crawled through the wire, avoided broken glass in a window or one of the massive front doors, and gone in. Now, although he says he's just curious and denies it, she suspects he is living there.

"You should see the place, Nana. Blues and greens, reds, yellows and oranges on the columns and archways, everything so big and beautiful it's like an explosion."

For a moment she thought the colors he saw were residue from the brain injury he'd suffered in Iraq, until he mentioned the graffiti, but it wasn't so much the colors that tipped her off how the station seduced him, it was the way his face seemed to take on a glow. She knows the thrill. It's the power of destruction he loves, as if a bomb had rocked the great door and windows, archways and columns into a trashed wasteland. How is it that the explosion in Iraq leads him to darkness while the shock wave knocking her and James from bed those many years ago, followed by a fireball broiling the sky, forever set her aflame?

A half hour it's been, maybe she should crawl into the building, but it's probably better to let her grandson know she's not stalking him, so she waits until a group of men duck through holes in the glass of the station's front doors. As Akin ghosts toward her, she steps from the car.

"I thought you might need a ride."

He climbs in the front seat and they drive home to Gilbert Street.

About two months ago her grandson shocked her when he showed up at her door. His mother had moved to Oakland, California when he was around two years old, one of the reasons she and her daughter had become so estranged. Sarah had wanted nothing to do with the city that every day reminded her of the father she never knew. "You're my only family, Nana," Akin said. "I want to be here for you." It was soon obvious, though, she needed to be here for him, a thirty-one-year-old man with no job. Refusing to stay with her, he told her he was living with friends.

Ella pours hot water from a kettle over her Sanka and carries her coffee cup and a bottle of beer to the kitchen table.

"I'm not supposed to drink, Nana."

"You stay in that train station you'll be drinking."

"It's the medicine."

"New Year's Eve, one's not going to hurt. I want you to come with me to the veterans' ceremony next week. If you show pride in your service and reconnect with the military, it will help you get back to normal."

You would think she had asked him to eat lunch with the Devil. He's ashamed to attend the city-sponsored celebration for war veterans on Belle Isle, with a special dedication to women who supported the troops during World War II. Since she's the only able-minded woman left in Detroit who built the big ships, she's been asked to give a little speech about what it was like in those days. What an honor.

"I can't, Nana. People still talk."

They do, particularly about men who walk away from war. She'd seen the newspaper accounts how members of his platoon had accused him of deserting after an explosion. They say he'd disappeared and Iraqi police found him days later wandering along a dry river bed. Although nothing came of an inquiry, and a doctor testified men who appear uninjured often suffer brain injuries from explosion shock waves, rumors keep circling. Well she knows something about rumors, how they feed on themselves.

Ella slices air with the sharp-boned edge of one hand. "The truth will come out."

That's what James Jackson said, after the Navy inquiry began. The look on his face, she'll never forget it, and to this day she doesn't know whether it was sheer hatred or something worse, a statement of fact. "Sooner or later the truth will come out, Ella. It always does." About ten years ago she thought she saw him in the post office on Springwells Street when a man in front of the line, next to the counter, held her gaze steady, and even then, some fifty plus years later, she turned away. When he walked past her, though, he was just another old man.

"The military was my life," Akin says. "I'm washed up. It's wrong."

"Wrong? You should know wrong."

After she and Arthur came home to Detroit, the factories wouldn't hire women welders, and she was supposed to stay home and feed her baby out of a bottle. Arthur was a good man and they'd had a good marriage; he accepted the child. Eventually she found work on the line at Fleetwood although it wasn't what she wanted. She wanted slabs of metal, steel beams, blast furnaces, and fire. Even Arthur, after thirty years on the line at Jefferson Assembly, didn't realize how useless he would become after retiring, but that's how it was in those days. Thirty years working, guaranteed out, the union had fought long and hard for the right to become nothing, and within a month after retiring he died of a heart attack. It amazes her that she herself is thirty years past retirement when it was a running joke among auto workers that after you retired you would drop dead the next day.

"When the war ended," she says, "everything changed suddenly. Like that!" She puts down her coffee cup and stops. Her grandson is staring at her. Coffee has spilled onto her hand and there's a chip on the bottom rim of her good china cup.

Cracks and booms everywhere, she shuffles to a front window to look out. It's midnight and her neighbors, whoever's left of them, are standing on the sidewalk shooting guns at the stars. Cherry bombs explode and rockets pierce holes in the curtain of night. She turns to wish Akin Happy New Year but the front door's wide open and her grandson's not there. Walking outside as quickly as her left hip will allow, she finds him standing on the sidewalk mesmerized by the fireworks, and when she reaches him he's clutching at the ends of his red flannel shirt, which is hanging out of his pants. He is weeping.

"Oh no, Akin, dear boy, everything will be all right."

When the neighborhood finally quiets and her grandson falls asleep in the second bedroom down the hall, she returns to her journal and flips open the gray cover to the first page. What did the class hope to achieve by keeping a journal, Felicia had asked her students. Write it down. But Ella could think of only two items: (a) explain what happened and (b) make things right. Lately she's started to see her days living in Port Chicago and working in Richmond in color, rather than in black and white, and last week in the Senior Center, while passing the open door of an art room on her way to her journaling class, a brown swirl on white canvas startled her. She saw James' hand on her skin.

What a time they'd had in an underground club over in Vallejo where there were other mixed couples, dancing, drinking, listening to music, always careful to avoid acknowledging each other if they happened to pass on Richmond's bustling streets. If she was with Arthur, who knew James only because he'd recognized him as one of the

mechanics in a Detroit garage where he took his car, and acknowledged him with a tilt of his head, or maybe a few words, she nodded, too. But that was about it.

Glancing through the window at the darkened Kelsey-Hayes plant, which the city is planning to demolish, she concentrates. Amazing how buildings, people, and even history change. How, depending upon your vantage point, the past is never still and memory is slippery and adjusts to the person you want to be. She'd been thinking about that for some time and after Akin arrived in Detroit, she acted and willed him the house. It's not worth much, but he can live in it or sell it to pay medical people to do whatever it is they do for traumatized people these days. Money, though, that comes and goes. Money is never enough. She doesn't know for sure her grandson didn't desert and imagines she never will, but one thing she will leave him is the certainty that James Jackson was right.

A hand on her thigh, a roar, the sky is on fire. Shredded curtains sway wildly, windows shatter, and Arthur's Navy black dress shoes glow orange. A second boom and she flies from bed, hits the wall, and falls to the floor. Pinned by a toppled dresser her ears hum, smoke and flame build massive columns in the night sky and Japanese warplanes fly through her fear. White light, screaming, the world is imploding, and yet she feels light. The weight on her legs disappears. Is Arthur pushing the bureau away? Oh yes, that's right, Arthur's on duty tonight. James searches her face with his fingers, lifts her from the floor and clings to her as they stumble past shards of glass to the front door. They jump off the porch, which no longer has steps. A mound of boards that used to be the porch of Umeko and Takeshi's bungalow rests on the ground, chimney bricks trail down the sloped roof, and Helen Richardson, who lives across the street, next to the Wilsons, appears in the doorframe of her house, which is missing the roof. Making their way along the street with a few of her neighbors – she can no longer recall who – they pass two upended car frames so twisted it's impossible to determine the make, pieces of metal from baby carriages and children's trikes, and splintered piles of wood that used to be houses.

"The ships," people are shouting.

Almost everyone's heading toward the port but smoke billows everywhere, occluding the sky, and forces them back as ghastly, orange clouds spirit through haze and fire in the distance, narrowing a slit between daylight and night.

"Where's the post office?" someone says. "Wasn't it over there?"

At one-thirty in the morning, it's a new year. A good time to reconcile an accounting of the disaster, to square up facts that have accumulated in her mind, particularly since she stopped working thirty years ago. To this day she wonders if she imagined what she witnessed or whether she created details that magnified the explosion to make it even more catastrophic than it was. As if such a thing were possible. Did she make up the fact that when she and James tried to reach the port a man stopped them and said *there ain't no port no* more? Or that one of the two ships men were loading with shells and bombs had been obliterated and the other had been blown out of the water?

It wouldn't have been possible to see those events, she later realized, since a firestorm was eating the sky and devouring structures and equipment that hadn't been annihilated. Especially the dead men, three hundred and twenty in all, many of their bodies were never found. And yet she's continued to embellish over the years, sometimes telling people over five hundred black men died or that close to a thousand people sustained injuries and she helped build one of the destroyed ships, the *S.S. E.A. Bryan*, when she hadn't.

It used to feel good to create facts about the explosion so devastating they would overwhelm crimes committed in the aftermath, make them shrink in comparison, seem small.

Ella resumes writing and considers what to include in the speech she plans to give Sunday, which she'll share with her journaling class. She will tell her audience it was considered a great honor to build liberty ships during the war while the men were off fighting, how after the war women were dismissed as if they were nothing, and what terrible injustices occurred such as sending the Matsumotos off to a camp and putting all those black men in jail.

When you think of it, which she does every day as she nears her end, the wrongs were enormous. It seemed to her within days of the explosion, even before all the body parts were located and the *Quinalt Victory* was discovered upside down in water five hundred yards away from the dock, rumors started. Enlisted men loading the ships with shells and bombs had sabotaged them, people said, and by the time the Navy Court of Inquiry began four days later, the rumors had fed on themselves and mushroomed. The enlisted laborers were of an inferior class. They had no experience handling live arms.

James was in the first group of men who refused to load ammunition onto a ship arriving at Mare Island three weeks later, and many of them were arrested. Although some of the surviving enlisted men had been evacuated to Camp Shoemaker in Oakland, he had been ordered to stay in Port Chicago to search for body parts. When he met her in a little motel next to the Horse Cow Bar in Vallejo, he seemed dazed.

"You've got to tell them, Ella," he said. "Tell them I was with you that night."

Shaken by the way his hands trembled and sweat dripped off his brow, she didn't respond. As perspiration dampened his heavy brows and made his skin glisten, he continually shifted toward a window and jumped when a door down the hall slammed. It wasn't so much how worried he was about being seen in the motel that rattled her as it was how fragmented he appeared, so pulled apart. He must have sensed how fearful she was about testifying, how the consequences might pull apart her own life, because that's when he told her, "Sooner or later the truth will come out, Ella. It always does."

She supposes her silence caused him doubt, just as some of her neighbors must have seen him on the walk to her house on those afternoons or nights when Arthur was working and never said anything. But someone betrayed her. After James was arrested, she heard people speculate how he was a leader in the mutiny because he wanted to sabotage the war effort and that's why he wasn't at the pier the night of the explosion. He was a deserter.

Years after the war she learned he did twelve years.

All of this Ella records. She writes compulsively as if her life depended upon the facts of who she once was and had strived to be and who she wants to become, which, when she considers herself at age eighteen and at the advanced age she is now, isn't all that much different. Finally, exhausted, at three-thirty in the morning she goes to bed, and when she wakes up, her grandson is gone.

> As perspiration dampened his heavy brows and made his skin glisten, he continually shifted toward a window and jumped when a door down the hall slammed.

When Akin doesn't visit her for five days, she toys with the idea of crawling through the hole in the barbed wire at the train station to confront him, yet hesitates and allows herself a glimmer of hope because not once before New Year's Eve had he slept at her house. She doesn't want to shock him although Sunday morning, surprise of surprises, Akin gives her a jolt. Banging the knocker on her door, the doorbell having given up close to fifteen years ago, he smiles at her nervously when she lets him inside.

"I won't go to the honoring ceremony, Nana, but I want you to know I'm proud of you."

This declaration of faith that should cause her joy, she's afraid she might cry. Her grandson leans over her. How much height has she lost over the years? Three inches? Four? She will not be diminished. Even after she dies she will not have anyone considering her small.

"I have important things to say. I want you to hear."

"No."

"Don't go to that train station anymore. I'm leaving you the house. When I'm gone –" he opens his mouth, but she silences him by raising a hand – "promise me you'll talk to medical people and stay here. And go through my papers. The deed and insurance documents are in the front of my journal. Promise you'll read it."

"I'm going to stay here tonight, Nana, but don't talk like that."

"Say it! You must say it!"

Her grandson looks slightly amused, and for some reason that gives her another small flicker of hope. "I do. I promise."

Using her good hip to hoist herself onto the middle rail, she leans forward. Staring downriver, the bridge lights come on, and so do the lights across the river in Windsor and at the Ren Cen and other skyscrapers.

Close to seventy people nod in solemn agreement when she tells them how women helped build the liberty ships in Richmond and how she lived in Port Chicago when the explosion destroyed so much. Not just the ships and port, nor the lives of men sent to prison, but something more important, the truth. Black enlisted men went to prison for mutiny while their white officers went free. Although the injustice of it all has long been acknowledged, she tells the audience, an assortment of veterans and government dignitaries, she's here to right wrongs. After her talk, people gather around tables to eat small cakes, puff pastries, and flakey-crust tartlets and drink tea and coffee as several people ask her questions and discuss their own service. Some even thank her.

When the sun dips low in the sky and casts orange light through the floor to ceiling windows, she leaves Flynn Pavilion and drives toward MacArthur Bridge, which spans the river from Belle Isle. When she reaches the bridge she stops, parks her car, gets out and walks to the rail to look at the city's skyline. Periodically turning toward her unlocked car, she doesn't want anyone stealing her purse on the front seat and tossing her identification, what's left of it, a library and medical insurance card, into a garbage bin. She will not be anonymous.

Using her good hip to hoist herself onto the middle rail, she leans forward. Staring downriver, the bridge lights come on, and so do the lights across the river in Windsor and at the Ren Cen and other skyscrapers. Sun slips down glass buildings, bronzes black water blue. She loves what this city once was. Zug Island blast furnaces throwing off flames, locomotives powering right into the factory at Rouge, and freighters carrying coke and iron ore pulling up to the docks. As silly as it sounds, she even liked the sight of Levy's old slag heap. Now look at the place. Three dollar cupcakes and five dollar coffee, for one cup, mind you, up near the university.

What a fine day this was. The audience was so appreciative and the elegant little pastries and flower arrangements on tables overlooking the river, something you might expect to find at classier gatherings. Sunlight poured through the windows the entire time she was there. Imagine! A sunny January day in Detroit although the temperature is frigid and ice flows drift on the river. Veterans and government officials will remember her, not that she told them everything, of course. She didn't mention, for example, the journal entry she wrote just this morning about the inquiry and the burdens some people carried years later.

And what would she have told them? That people were different during the war so wrongs were committed? You don't share a journal with strangers. Anyway, these days it's all public record, so if people want to know more about the explosion and its aftermath they can look it up on their computers or go to the library. Strangers don't need to know details that have stayed with her all these years, the questions the officer at the inquiry asked, and how she tilted toward him and focused on his medals to still her anxiety. The man's uniform was deep blue, his medals bright brass.

"Mr. Jackson claims he was with you."

"I can't help what he claims."

"He was not at your house that night?"

"No."

"Well, which is it? Was he there or was he not?"

"No Sir. He was not."

To this day she recalls how she tried to narrow the distance between her and the officer, as if by angling into him she could project confidence and weaken his doubts, and apparently it worked. People, when confronted with difficult choices, will rationalize the truth. Ella leans over the rail toward the Navy man, stretches toward bronze and blue, and as the sun falls away and the Detroit River blackens, she closes the distance between them, and the day's light goes out.

Spruce Island, ME, 2015

Ron Johns
photograph

Laura Apol

Midwinter, My Mother

After I left the cemetery, I drove —
east to the ocean, seven states away.
The journey was slow, weighted as I was,
myself in my arms. I thought

she would go with me. I thought
she would stay with me in my sleep. Instead,
I dreamed of fallen horses, ruins of battle.
— those useless limbs, those dying

horses' eyes. I walked on the beach
until I understood: how water and time
grind down the world, hand us our cartilage,
broken. Hand us our bones.

She'd known, in the end, because
I had to tell her. I stood by her bed,
kept my mind on the moon, the clear
winter light. I wanted

to hear her say she loved me,
and I pretended she did, pretended
I heard those words in the waves —
above shards of shell, fish bones picked clean.

#18

Rick Johns
acrylic and graphite on wood panel

On Kindness

Laura S. Distelheim

Black is how I remember it. The kind of black that won't meet your eyes. That looks away, unimpressed, no matter how fervently you beg it for mercy. Why I was there — alone in my stalled car, by the side of a road whose name I wasn't sure of — and where I was en route to or from, I've since forgotten. This was long ago. But the vacuum around me, the moonlessness and the mercilessness, the *hoos* and whistles and hisses and screeches of nocturnal secrets, and the terror, I remember. *That* I remember, and this: That, when my fingers finally managed to stutter out the number of my emergency roadside assistance service on my cell phone, the voice that bloomed on the other end of the line answered with, *Are you somewhere that you feel safe?* Not a question at all, but a life vest, and I slipped my arms into its kindness and buckled it around me and zipped it straight up to my chin.

I once dated a man whose father had warned him against being too kind for fear that it would make people perceive him as weak. I hated hearing that when he told it to me, less for what it said about him (because, despite his father's best efforts, he was, in fact, kind), than for what it said about kindness itself. I, for one, know that it gets a bad rap. Want me to show you? Come with me: Back to a morning just over a year ago, at a hospital fifty or so miles from my home, where I have gone to be subjected to a test I don't want to take, ordered by a doctor I don't want to know, for a chronic illness I don't want to have, that has cast me in a life story I don't want to be mine, all of which I have muttered and ranted about in perfect duet with Springsteen, set to full blare (*No retreat, baby, noooooooooooooooooo surrENder),* for nearly the entire hour and a half it has taken me to get here, my hands vised on the steering wheel as I centimetered my way through a clot of city traffic (starting, stopping, starting, stopping, getting lost, retracing, starting, stopping) so that, by the time I pull up to its entrance, I'm dragging the full fury and frustration inflamed by decades of battles — of lying in antiseptic halls on pushed-to-the-side gurneys, waiting for radiology to call me in; of being the g.i.case / infectious disease case / ortho case / neurology case / rheumatology case / pre-op case / post-op case in room 512 /room 780 / room 432 / room 285; of wearing green gown after green gown after green gown that turned my face anonymous — behind me like a

ruckus of tin cans that are clattering from my fender.

It's no wonder the valet parking attendant looks up, startled, when I stagger out of the driver's seat. "I have a handicapped placard," I tell him as I give him my keys, because that's what the sign on the wall behind him instructs me to do. If I have one. Which I do. Which I do and do and do. Which I DO. It isn't his fault. I know that *none* of this is his fault, but the words come out more like an accusation than an announcement anyway, which may explain why he repeats them as a question; "You have a handicapped placard?"

He can see this any way he wants to. He can see this as a race thing, a class thing, a socioeconomic thing, a simple rudeness thing. But instead his voice is full of knowing exactly what that means — what the words mean and what the way I've said them means — and the look he gives me is a hand that I hold onto for the rest of the day. Weak? I don't think so. In fact, here it is, his gentle gesture, more than a full year later, still going strong.

Kindness has a way of doing that — of living on long past the culmination of its actual lifespan. And, unfortunately, lack of kindness shares the same skill. "What I regret most in my life are failures of kindness," the author, George Saunders, told the 2013 graduating class of Syracuse University when delivering its convocation speech and, hearing him repeat that statement in a recent television interview, I was immediately transported to a twilit moment nearly three decades ago on a sidewalk in Cambridge, Massachusetts, where I was a few months into my first year as a student at Harvard Law School.

This wasn't my first time as a first year grad student. Three years earlier, still reeling from my most recent ten rounds with my illness, I'd started as a student at another school, where it had felt, from the moment I set foot on its campus, as if I were wearing my life like a misbuttoned coat. No matter how much I yanked at its collar or tugged at its sleeves, that life continued to hang crooked from my shoulders, until finally, at the end of that year, I'd unbuttoned it, tore it off, and left it in a puddle behind me. So that when, two years of regathering strength and regaining direction later, I was accepted to HLS, it had felt less like I'd been invited back to grad school than that I'd been handed yet another life to try on, and I had started out the year there holding my breath, not daring to believe that it would fit me.

It did. Slipped onto my skin and fell into place around my bones as if it had been custom tailored for me, which is why, there I was, walking down the sidewalk at that twilit moment, surfing the crest of a wave of relief. And there, heading toward me, half a block or so away, was a classmate whom I'd never talked to but had seen across the room in several of my classes, unmistakably tugging at the sleeves and yanking at the collar of his own life. Something about the way his shoulders were hunched, about the way his hands were in his pockets, about the way his head was cocked and his smile was almost apologetic in the moment when our eyes met as we passed one another, gave me the impulse to reach out and touch his arm and ask if he wanted to stop in somewhere, maybe, have a cup of coffee maybe, and, you know, just talk. But I didn't because, well, I didn't.

> It isn't easy, kindness. You might even say that it can be perilous because, let's face it: It can backfire big time.

I didn't because I wasn't the kind of woman who did those types of things, who moved through the world with that kind of easy grace. I had always been the sick girl who was frequently missing long stretches of school and then returning on crutches or with a new set of scars or twenty pounds thinner. And even though I thought I had felt something – a sadness, a *please*, a *come get me* – when our eyes met as we passed each other at that moment, I hadn't trusted it. *What if I'm wrong?* or *What if he doesn't want to?* or *What if he thinks it's strange of me to ask?* is what I'd busied myself with thinking as I'd moved on past him on the sidewalk and so no, I hadn't trusted it.

Not at first, anyway. Not until the next day, when I'd finally gathered my courage and had gone to his dorm room and had found the door ajar and the room in-your-face empty. Not just empty, but emptied out. His closet door open, its hangers naked and askew, his bed stripped and his desktop swept clean, the chair – the chair where I had imagined I'd find him sitting, in all the scenarios I'd practiced in my head on my way over – set back at an angle, vacant. Dropped out, a guy who lived in a room across the hall told me when I knocked on his door to ask. Dropped out with no forwarding address.

It's still with me, that moment on the sidewalk when I passed him and didn't reach out. It's still with me all these years later, having trailed me through the decades of twilights that have bloomed and waned since that one. It doesn't come to me often, but there are still moments, maybe especially in winter if I'm walking down a sidewalk where the hills of snow that have been plowed up against the curb are soaking the dusk into their folds until they have blued, when I'll find myself imag-

ining for an instant that there he is, coming toward me, and knowing all over again that no, no he's not. That that chance to be kind is long past and I can never get it back. Can never do it over, do it right, reach out, touch his arm, say those words, let him know that he's not the only one.

So yes, I'm right there beside George Saunders in counting failures of kindness among my greatest regrets. Which is why I've done my best ever since that time to try to avoid them. But even as I have, I've understood the self who didn't stop on the sidewalk that day. It isn't easy, kindness. You might even say that it can be perilous because, let's face it: It can backfire big time. I've learned that lesson the hard way, too. A few years ago, for instance, a few months after a friend whom I had met through a writers' group lost her battle with breast cancer – a battle that had never prevented her from reaching out to support me in either my writing or my own never ending struggles with my health – when I decided to drop a note to her widower, whom I had met just once, briefly.

Because I had often heard it said that one of the most excruciating times after the loss of a loved one is the period when the memorials have been held and the condolence cards have been sent and the rest of the world returns to tending its own business while you're left alone with your grief, I decided that letting him know that I was still missing my friend, too, would be a way of paying forward her many kindnesses to me. One last chance to thank her, I thought as I dropped the note in the mailbox, and then promptly forgot that I'd sent it. Until I received his note back. He was missing her too, he said. In fact, he would *always* be missing her, he said. And furthermore, he said, he wanted to make it perfectly clear that she was the *only* woman he would *ever* be interested in missing in the way that he missed her.

Okay then. I still cringe every time I recall the words in his note, and although I spent some time contemplating writing again with an "Ohmygod, you completely misunderstood!," I soon came to realize that the more I protested, the less he would believe me, and so I had no choice but to let it go. Or, at least, had no choice but to *try* to let it go, but what I've discovered in the time since is that that's not so easily done. Which is why, while – despite the periodic bouts of cringing – I still don't regret that I reached out in kindness that time, what I do regret is that his response to that kindness has kept me, on occasion, from reaching out again.

On *this* occasion, for instance: It's a year or so later and I'm sitting in my car at Lake Michigan's edge in the hour before sunrise, where I often start my day and where, on this particular day, Beethoven's ghost is playing "Moonlight Sonata" all over the beach. The world is swathed in black velvet, with little light in the sky, save for a crescent moon that's sending a circle of nacre down to float upon the water. Lightning is flashing every now and then – postcards mailed home from last night's storm, now that it's moved on – and after a time, the horizon exhales its first breath of light, until, suddenly, the entire span of the sky is cast in a delft shade of blue with patches of pearl drifting across it. The stage has been set for a clarinet or trumpet solo, smooth and seamless and shivering with grace, and I slip

Gershwin's "Summertime" into my CD player just so that I can keep playing and replaying its opening lament.

It had been summertime, I find myself thinking then, when I first met David and Ken. A summertime a few years after my law school graduation, not long after my illness had ambushed me once again, knocking the breath out of me to the point where I'd had to leave the career and the relationship and the life I'd been creating and move back to my parents' home. David had been a close friend of my brother-in-law's since college and, when my parents had learned that he and his partner, Ken, were in town for a visit, they'd invited them, along with my sister and brother-in-law, to dinner. A dinner which I attended bedecked head to toe in embarrassment and shame.

Since returning to my home town, I had rarely left the house, even during the short periods when I had enough energy to do so, so filled was I with dread at the prospect of running into someone I knew, or had once known, and finding myself coming up against the inevitable, "So, what are you *up* to these days?" Despite all that my head understood about the enormity of the health battle I was waging, and had been waging for most of my life, my heart continued to insist that "FAILURE" was stamped across my forehead. After all, was how my internal argument went, I had only one thing on my To Do list – "Recover!" – and I couldn't seem to get it checked off.

It seemed the worst kind of defeat to me to have made it all the way to my law school graduation through an obstacle course of two operations and three additional hospital stays only to now find myself back where I'd started. And that sense of defeat was only heightened by the fact that the illness I was battling was so poorly understood that those few friends I still had (those who hadn't already found all of this too difficult to relate to and quietly tiptoed away) had long ago given up on trying to understand it. Which is why *I* had long ago given up on trying to explain it. "I'mfinehowareyou?" I would say when cornered, always racing to deflect the focus of any conversation away from myself, so that it sometimes felt as if I were engaged in a perpetual game of hot potato.

But David and Ken, I rapidly discovered that night, weren't playing. "No, not yet," each of them said when I tried to toss the potato to him, and "Really, *tell* us," they said, and then they asked everything it appeared they genuinely wanted to know, about how it felt, this illness, and not only this illness, but the fear that was its Siamese twin, and about what it had done to my body, and not only to my body, but to my life as well. And then we talked about HIV and AIDS and how like that battle mine was, and by the time the meal was through and my mother was serving the sorbet, I had stopped being lost.

It was summertime, the evening of that dinner, but that's not why I find myself thinking of it as I sit on the beach on that years later morning, with a new day developing like a photograph in a dark room around me and Gershwin's composition shimmying and reshimmying its way to a moan. The reason I find myself thinking of that dinner at that moment is that I learned just last night that David has lost Ken

to his own lengthy health battle. And so, as soon as I return home from the beach, I sit down and write him an e-mail, telling him how much their kindness at that dinner had meant to me, and how much it had stayed with me since, and how much I wish I had told both of them that at the time, and how much he is in my thoughts now. I spend a long time composing that e-mail and then I revise it and reread it and reread it once more before guiding my cursor toward "Send." Which is when the thought of my friend's widower's response to that other note resurfaces in my mind and I find myself clicking on "Save to Drafts" instead.

There is no danger, of course, of David's misconstruing my intentions in the way that my friend's widower had, and yet that startling response to what I had thought was a simple gesture of kindness has made me so second guess myself that I find myself hesitating to send this e-mail anyway. I have seen David and Ken only once since that long ago dinner, this time at my niece's Bar Mitzvah party shortly before Ken died, where, in the middle of a room pulsing with gyrating teenagers and Klezmering musicians and to-and-fro-ing waiters and shouting-to-be-heard adults, we'd managed only a cursory chat. Will David think it's strange for me to still be remembering that dinner? I find myself wondering. Will he think it startling even? I don't know.

I can't be sure, because one thing I do know for sure is that my years of being so frequently side lined, of looking at the world so often from a distance, have left me seeing it differently from the way that most people do. Have left me changed to the point where it sometimes feels as if I've developed something akin to the bioluminescence which I recently read is used by the Western Grebe, a bird that lives alongside large, deep lakes in the western half of North America, to enable it to forage for food at night, locating even the most elusive of fish in total darkness. It appears that I've developed a type of bioluminescence for locating places where kindness is needed.

Maybe that was unavoidable, given how many times I've been in need of kindness myself, and, in any case, I'm at peace with having developed this somewhat odd faculty, but if that cringe-inducing response I'd received had taught me anything, it was that it's a faculty that I have to think twice before using if I don't want to leave a trail of raised eyebrows in my wake. And so I leave it there, in drafts, that e-mail to David, a frozen embryo of kindness that, as it turns out, I will never allow to develop to full term. I have regretted that decision every time I have thought of it since.

What I don't regret is having acquired that bioluminescence, because it enables me to zero in not only on the places where kindness might be needed, but on those where it's being committed as well, even when the person committing it is doing so hidden from view. Take a man I'll call Gus, for instance. A small, prickly man who is the manager of the grocery store at the center of my town, he's prone to long, gloomy silences and short, staccato bursts of speech that are designed to cement his reputation as a curmudgeon. "How was your vacation?" I once asked him after he'd been away for a couple of weeks, and "Terrible," he answered. Pause. "It ended." And: "How did